AF718679

Medicinal plants of the Andes and the Amazon

The magic and medicinal flora of Northern Peru

MEDICINAL PLANTS OF THE ANDES AND THE AMAZON

The magic and medicinal flora of Northern Peru

Rainer W. Bussmann & Douglas Sharon

©Copyright ***Rainer W. Bussmann & Douglas Sharon***

Copyrighted material. Any reproduction, storage in databases system or trasmission of any part of this publication, including throug than employee, in any electronic, mechanical, photocopied or recorded form etc. is prohibited without the autor's permission.

ISBN-10:0-9960231-2-7
ISBN-13:978-0-9960231-2-2

$29.99
ISBN 978-0-9960231-2-2
52999>
9 780996 023122

1ª ed., setiembre 2015
Hecho el Depósito Legal en la Biblioteca Nacional del Perú Nº 2015-11583
Edited by Rainer W. Bussmann & Douglas Sharon
Graphic design and editorial support: Carlos Vega Ocaña
Printed in Graficart srl
San Martín 375 Trujillo - Perú
Telf: 044297481

PRINTED IN PERU

MEDICINAL PLANTS OF THE ANDES AND THE AMAZON

The magic and medicinal flora of Northern Peru

Rainer W. Bussmann – Douglas Sharon

DISCLAIMER

The statements in this book are a compilation of Northern Peruvian traditional medicine and belief, gathered from interviews with traditional healers and market venders, in order to document and preserve this traditional knowledge. The information has been recorded as presented by the participants interviewed.

This book does not purport to contain, nor is it intended to be, any kind of recommendation or self-treatment guide for the use of plants or traditional practices. Some of the species reported are quite toxic or could cause harm when used inappropriately. Even plants that are commonly consumed as food and reported to be generally recognized as safe may have adverse effects, including drug interactions and hypersensitivity in some individuals.

The authors have checked with sources believed to be reliable to confirm the accuracy and completeness of the content of this book. However, the authors, editors, copyright holders, and the publisher disclaim all warranties, expressed and implied, to the extent permitted by law, that the contents are in every respect accurate and complete, and they are not responsible for errors, omissions, or any consequences from the application of this book's contents. Nothing in this book should be construed to represent an attempt to diagnose, prescribe, or administer in any manner any physical or psychological ailments or conditions, nor should the information in this book be used in place of qualified medicinal advice and care.

Although the information provided in this book has been presented to promote education and scientific research, we realize that elements of this work have the potential to contribute to commercialization. All work for this book was carried out under the rules of the Nagoya Protocol regarding access to genetic resources and fair and equitable sharing of benefits arising from their Utilization under the Convention on Biological Diversity. The right of ownership of this a knowledge is maintained by the traditional healers and venders who were interviewed and who have developed their knowledge through centuries of experimentation and experience. Any commercial use of the information presented in this publication requires the prior consen of informants and communities and an agreement on the distribution of benefits.

For Ivan Gayler, who's generosity started our work on Andean ethnobotany in 1995

Traditional and Complementary - Alternative Medicine

Traditional Medicine is used globally and has a rapidly growing economic importance. In developing countries, Traditional Medicine is often the only accessible and affordable treatment available. In Uganda, for instance, the ratio of traditional practitioners to the population is between 1:200 and 1:400, while the availability of Western doctors is typically 1:20,000 or less. Moreover, doctors are mostly located in cities and other urban areas, and are therefore inaccessible to rural populations. In Africa, up to 80% of the population uses Traditional Medicine as the primary healthcare system. In Latin America, the WHO Regional Office for the Americas (PAHO) reports that 71% of the population in Chile and 40% of the population in Colombia have used Traditional Medicine. In many Asian countries Traditional Medicine is widely used, even though Western medicine is often readily available. In Japan, 60–70% of allopathic doctors prescribe traditional medicines for their patients. In China, Traditional Medicine accounts for about 40% of all healthcare, and is used to treat roughly 200 million patients annually. The number of visits to providers of Complementary-Alternative Medicine (CAM) now exceeds by far the number of visits to all primary care physicians in the US (WHO 1999, 2002; WHO/PAHO1999).

Complementary-Alternative Medicine is becoming more and more popular in many developed countries. Forty-eight percent of the population in Australia, 70% in Canada, 42% in the US, 38% in Belgium and 75% in France, have used Complementary-Alternative Medicine at least once (WHO 1998; Fisher and Ward 1994; Health Canada 2001). A survey of 610 Swiss doctors showed that 46% had used some form of CAM, mainly homeopathy and acupuncture. (Domenighetti et al. 2000). In the United Kingdom, almost 40% of all general allopathic practitioners offer some form of CAM referral or access (Zollman and Vickers 2000). In the US, a national survey reported the use of at least 1 of 16 alternative therapies increased from 34% in 1990 to 42% in 1997 (Eisenberg 1998; UNCTD 2000).

The expenses for the use of Traditional and Complementary-Alternative Medicine are exponentially growing in many parts of the world. In Malaysia, an estimated US $ 500 million is spent annually on Traditional Medicine, compared to about US $ 300 million on allopathic medicine. The 1997 out-of-pocket Complementary-Alternative Medicine expenditure was estimated at US $ 2,700 million in the USA. In Australia, Canada, and the United Kingdom, annual Complementary-Alternative Medicine expenditure is estimated at US $ 80 million, US $ 2,400 million and US $ 2,300 million, respectively. The world market for herbal medicines based on traditional knowledge was estimated at US $ 60,000 million in the late 1990s (Breevort 1998). A decade later it was around US $ 60 billion (Tilburt and Kaptchuk 2008) with estimates for 2015 at US $ 90 billion (Global Industry Analysts 2012). The sales of herbs and herbal nutritional supplements in the US increased 101% between May 1996 and May 1998. The most popular herbal products included Ginseng (*Ginkgo biloba*), Garlic (*Allium sativum*), *Echinacea* spp, and St. John's Wort (*Hypericum perforatum*) (Breevort 1998).

Traditional and Complementary-Alternative Medicine are gaining more and more respect by national governments and health providers. Peru's National Program in Complementary Medicine and the Pan American Health Organization recently compared

Complementary Medicine to Allopathic Medicine in clinics and hospitals operating within the Peruvian Social Security System (EsSalud). A total of 339 patients — 170 being treated with Complementary-Alternative Medicine and 169 with allopathic medicine — were followed for one year. Treatments for osteoarthritis; back pain; neurosis; asthma; peptic acid disease; tension and migraine headache; and obesity were analyzed. The results, with 95% significance, showed that the cost of using Complementary-Alternative Medicine was less than the cost of Western therapy. In addition, for each of the criteria evaluated — clinical efficacy, user satisfaction, and future risk reduction — Complementary-Alternative Medicine's efficacy was higher than that of conventional treatments, including fewer side effects, higher perception of efficacy by both the patients and the clinics, and a 53–63% higher cost efficiency of Complementary-Alternative Medicine over that of conventional treatments for the selected conditions (EsSalud/ OPS 2000).

According to WHO (2002), the most important challenges for Traditional Medicine/ Complementary-Alternative Medicine for the next years are:

- *Research into safe and effective Traditional Medicine and Complementary Alternative Medicine treatments for diseases that represent the greatest burden, particularly among poorer populations.*
- *Recognition of the role of Traditional Medicine practitioners in providing healthcare in developing countries.*
- *Optimized and upgraded skills of Traditional Medicine practitioners in developing countries.*
- *Protection and preservation of the knowledge of Indigenous Traditional Medicine.*
- *Sustainable cultivation of medicinal plants.*
- *Reliable information for consumers on the proper use of Traditional Medicine and Complementary-Alternative Medicine therapies and products.*

Dr. Manuel Fernández (2009) national sub-director of Peru's Instituto Nacional de Medicina Tradicional (INMETRA) in the 90s delineated problems related to the production of phyto-pharmaceuticals in Peru:

- *Lack of national policies.*
- *Absence of state and local policies that include medicinal plants.*
- *Lack of support by the state.*
- *Lack of support from the medical establishment.*
- *Ignorance of the benefits of the phyto-pharmaceutical industry.*
- *Limited human and technical resources.*
- *Lack of technical knowledge for the production of herbal products.*
- *Ignorance of methods and processes of quality control and standardization.*

- *Problems in obtaining quality materia prima in adequate quantities and predatory collecting.*
- *Absence of conservation policies implementing cultivation of herbals under best conditions.*
- *Limited research in ethnobotany, agrotechnology, pharmacy, and therapeutic validation.*
- *Lack of legal parameters for sanitary controls and commercialization of herbal products.*
- *Vested interests of the pharmaceutical industry minimizing the importance of herbals which are not the product of their own research and development.*

Dr. Fernández also discussed a decreasing trend in Latin American consumption of medicinal products from 8% of global consumption in 1980 to 5% in 1990. He attributed this reduction to decreased government distribution of free medicines to the poor, concentrated wealth in a few hands, and increased poverty. Another factor was the fact that developed nations spend a much higher percentage of GDP on medicines (6-8%) than developing nations (1-2%) where it is estimated that two -thirds of medicines purchased are paid for by the patients. And per capita spending is much higher in developed nations compared with developing countries, e.g., Japan: US $ 276; Germany: US $ 148; USA: US $128; Argentina: US $ 42; Uruguay US $ 40; Paraguay: US $18; Brazil: US $ 10.5; Bolivia US $ 4. There were no figures for Peru, but it was estimated to be slightly above the amount for Bolivia. Overall, it was estimated that 50% of the population of Latin America had little or no access to medicines and that a large portion of these people used medicinal plants.

An innovative response to the challenges listed above has been developed by the Centro de Medicina Andina (CMA) founded in Cuzco in 1984 as an autonomous branch of the Instituto Pastoral Andino (IPA). Started by Catholic healthcare workers with extensive experience working in Quechua communities, CMA's pragmatic methodology involves "mutual training" between healthcare professionals, community health promoters, *curanderos*, and midwives. For them the rhetorical question is: "Who knows all of the richness of Andean medicine better than the peasant himself, the specialist-practitioner of this medicine?"

Objectives of the Centro up to 1992 were: "1) Advance a health system favoring the majority of the community where Natural-Popular Medicine and modern medicine are complementary. 2) By means of study and application of Natural-Popular Medicine, create a scientific basis for its development." Revised objectives since 1992 are: "1) Value and rescue Andean Medicine in order to contribute to better utilization and recognition within a system of alternative health available to a majority of the population. 2) Investigate, experiment, and disseminate the experiences and knowledge of Andean Medicine. 3) Encourage debate, exchange, and coordination between people and institutions working in the field of Natural-Popular Medicine. 4) Rescue Andean food plants to improve food consumption."

CMA's programs include: 1) *Education*—"Peasant to peasant" training of community health promoters and women's groups in cooperation with local universities and the Ministry of Health. 2) *Medicine and Medical Anthropology*—Epidemiological and regional health-status diagnoses, evaluation of traditional therapies, and ethnography and publication of aspects of Andean culture and "cosmovision." 3) *Ethnobotany and Phytotherapy*—Collection and identification of 3,740 plants and development of an herbarium and certified laboratory leading to the production and commercialization of six natural medicines (Garrafa R & R: 2009).

Another innovative project was organized as part of the local primary health care program in Puquio (Ayacucho) in 1997 by P. Dieter Wacker (MSC), President of Asociación Huitco. It includes a botanical garden; an herbarium; drying and storage areas; processing equipment; supply chains (regional and international) for ointments made from *huitco, camasol, molle* and eucalyptus as well as eucalyptus syrup; quality and safety certification, ethnobotanical research, and educational programs all administered by local personnel (Pietrellini 2007: 3-5).

The present study, financed through the "MHIRT," attempts to address some of the issues discussed above. MIRT (Minority International Research and Training) renamed MHIRT (Minority Health and Health Disparities International Research and Training), is funded by Grant G0000613 from the National Institutes of Health. Initially it was administered by the Fogarty International Center for Advanced Studies in Washington, D.C.; currently by the Health and Health Disparities Program. MHIRT-Peru is coordinated by San Diego State University (SDSU, Dena Plemmons, Ph.D) in cooperation with the Missouri Botanical Garden (MBG), the University of California-San Diego (UCSD), the University of California-Berkeley (UCB), the University at Buffalo (SUNY Buffalo), and Linfield College — in the US – and the Universidad Privada Antenor Orrego (UPAO), the Universidad Nacional de Trujillo (UNT), the Clínica Anticona Trujillo (CAT), and the Centro de Atención en Medicina Complementaria (CAMEC)-EsSalud Trujillo - in Peru.

The primary focus of this project has been the ethnobotany of medicinal plants used on the north coast of Peru. Northern Peru represents the upper sector of the "health axis" of the ancient Central Andean cultural area stretching from southern Ecuador to northern Bolivia (Camino 1992). The traditional use of medicinal plants in this region, which encompasses in particular the Departments of Piura, Lambayeque, La Libertad, Cajamarca, Amazonas, and San Martin possibly dates as far back as the first millennium B.C. (north coastal Cupisnique Culture) or at least to the Moche period (A.D. 100-800).

Precedents for this study have been established by early colonial period chroniclers (Monardes 1574; Acosta 1590; Cobo 1653); the plant collections (293 plants in crates 11 and 12 of 24) of Bishop Baltasar Jaime Martínez Compañón sent to the Palacio Real de Madrid along with cultural materials in 1789 under the title *Trujillo del Perú* in 9 illustrated volumes (Schjellerup 2009; Sharon and Bussmann 2006); the work of Ruiz and Pavón (1798, 1799, 1802) and Italian naturalist Antonio Raimondi (1857); ethnoarchaeological analysis of the psychedelic San Pedro cactus (Sharon 2000); *curandera* depictions in Moche ceramics (Glass-Coffin, Sharon, and Uceda 2004); and research on the medicinal plants of Southern Ecuador (Béjar, Bussmann, Roa, and Sharon 1997, 2001; Bussmann and Sharon 2006a, 2007a).

Fieldwork for the present study started in the markets of Trujillo (Mayorista and Hermelindas) and Chiclayo (Modelo and Moshoqueque) in 2001. In this book we provide the North Peruvian data and illustrations from the ethnobotanical database (Skoczen and Bussmann 2005) of 510 medicinal plants organized under the headings: scientific name, vernacular name, plant parts used, administration, preparation, and use. (Bussmann and Sharon 2006b, 2007b).

Antecedents - Medicinal Plant Research and Traditional Medicine in Peru

Containing 84 of the 107 eco-regions of the world, in 1993, it was estimated that Peru had 17,143 taxa of spermatophytes in 2,485 genera and 224 families (Brako and Zarucchi 1993). León et al. (2006) found 18,650 taxa. It is thought that only 60% of the Peruvian flora has been studied, with 1,400 medicinal species described (Brack 2004).

The importance of biodiversity for the Peruvian economy is enormous since 25% of all exports are living species, the uses of which are essential to local populations in terms of firewood, meat, lumber, medicinal plants, and many other products. Of particular importance are vegetal species, with 5,000 plants applied in 49 different uses. Of the 5,000 plants in use some 4,400 are native; only 600 are introduced. The majority of useful native species are not cultivated; only 222 can be considered to be domesticated or semi-domesticated (Brack 1999).

Transculturation is resulting in an enormous loss of traditional knowledge of great value to the science and technology of Peru. The flora of the country represents 10% of the world's total, of which 30% is endemic. Peru is the fifth country in the world in number of plant species with known properties utilized by the population (4,400 species); it is the first in domesticated native species (182) (Brack 2004).

In all Peruvian ethnic groups, plant knowledge is invaluable because it reinforces national identity and values which are being lost in the complementary processes of modernization and globalization. In the current situation the emerging recognition and incipient application of these resources and associated knowledge emphatically underscores the critical need for ethnobotanical research in light of the following facts:

- Absorption and devaluing of native culture due to modernization and globalization.
- At the same time, recuperation and revaluing traditional knowledge of Peruvian flora.
- Emerging "first world" awareness of the therapeutic potential of medicinal plants.
- Recent ethnobotanical research by a growing number of Peruvian scholars (Alban 1985).

In *Sinopsis histórica de la Etnobotánica en el Perú,* La Torre and Alban (2005) outline the history of formal floristic studies in Peru starting in the late 1700s with the work of Hipólito Ruiz, José Pavón and Joseph Dombey followed by Alexander von Humbolt, Aime Bonpland, and Antonio Raimondi in the nineteenth century. Twentieth-century scholars included: Fortunato Herrera, Hermilio Valdizán and Angel Maldonado, Juan Lastres, Jaroslav Soukop, Julio López and Irma Kiyán, Felipe Chavez, Emma Cerrate de Ferreira, James Duke and Rodolfo Vásquez, and Joaquina Alban. However, it was John Harshberger who used the term ethnobotany for the first time in Peru in 1896 while Juana Infantes actually established the discipline at the Universidad Nacional Mayor de San Marcos in 1945.

Considerable progress has been made in the overall taxonomic treatment of the flora of Peru over the last few decades (Brako and Zarucchi 1993). However, while the Amazon rainforests have received a great deal of scientific attention, the mountain forests and remote highland areas are still relatively unexplored. Until the late 1990s little work had been done on vegetation structure, ecology, and ethnobotany in the mountain forests and coastal areas of the North. In spite of the fact that this region is the core of what Peruvian anthropologist Lupe Camino (1922) calls the "health axis" of Central Andean ethnomedicine, little ethnobotanical and ethnomedical research had been published on the rich flora found here.

Early ethnobotanically oriented studies focused mainly on the famous "magical" and "mind altering" flora of Peru. A first study on "cimora" - another vernacular name for the San Pedro cactus, dates back to the 1940's (Cruz Sánchez 1948). The first detailed study of a hallucinogen in Peru focused on the San Pedro cactus (*Echinopsis pachanoi*) (Dobkin de Rios 1968, 1969). A variety of works including some on the "Daturas" (*Brugmansia* spp.) followed (Bristol 1969; Crosby and McLaughlin 1973; Dobkin de Rios 1977, 1980; Pummangura et al. 1982). Coca (*Erythroxylon coca*) also attracted early scientific attention (Martin 1970; Naranjo 1981; Plowman 1981, 1984 a, b) as did the Amazonian Ayahuasca (*Banisteriopsis caapi*) (Rivier and Lindgren 1971; McKenna, Luna and Towers 1986; Schultes and Raffauf 1992; Bianchi and Samorini 1993). Chiappe, Lemlij and Millones (1985) were the first to attempt an overview on the use of hallucinogens in shamanistic practices in Peru. More comprehensive accounts followed (Alarco de Zandra 1988; Cabieses 1990; Schultes and Hofmann 1992; Schultes and Raffauf 1990).

In his classical study of Uña de Gato, Peru's leading advocate for Traditional Medicine and founding director of the Instituto Nacional de Medicina Tradicional del Ministerio de Salud, Fernando Cabieses (2000) pointed out that the work of the Peruvian scholars Hermilio Valdizán and Ángel Maldonado (1922) was the pioneering effort in studying Traditional Medicine, leading to the emergence of medical anthropology nearly five decades later. In the interim the botanical exploration of Peruvian flora, and medicinal plants in particular, included studies by Yakovleff and Larco Herrera (1935); Weberbauer (1945); Towle (1961); and Valdivia (1975). Most authors (Larco Herrera 1945; Herrera 1941; Lira 1985; Soukup 1970, 1987; Franquemont, Plowman et al. 1990; De Ferreyra 1981) focused on Quechua herbalism of the Cuzco area. Other comprehensive studies centered on the border region of Peru and Bolivia around Lake Titicaca (Girault 1987; Bastien 1987; Roersch 1994; Macia el al. 2005) and the Amazon (Prance 1972; Vásquez 1989; Ritter 1990; Duke 1994; Jovel 2005), Cabieses (1993) wrote a major tract on Traditional Medicine while Ugent and Ochoa (2006) and Fernández H. and Rodriguez R. (2007) summarized pre-Hispanic ethnobotany. Detailed studies of Uña de Gato (Obregón 1996), Maca (Cabieses (1997), and Sangre del Grado (Meza 1999) were also carried out.

Initially, Northern Peru was in the shadow of other areas studied, attracting little scholarly attention until late in the twentieth century. Much work has now been done in the mountain forests of the north, including studies of the floristic diversity and negative state of conservation of the Department of Cajamarca and part of the Department of Amazonas (Sagástegui et al. 1999 and 2003) while the National Museum of Denmark has placed the ethnobotany of parts of the departments of Amazonas and San Martín in cultural and ecological context (Schjellerup et al. 1999, 2003, 2005). In the Ayabaca District of the Department of Piura, De Feo (1992, 2003) collected 46 medicinal and magical plant species and their uses particularly in relation to divination practices. In the Callejón de Huaylas (Ancash), Hammond et al. (1998) e conducted a survey of traditional medicinal plants complemented by the collection of 178 medicinal plants and their uses (Gonzales de la Cruz et al. 2014). In Bolivar Province (La Libertad) Monigatti, Bussmann and Weckerle (2012) collected 2,776 plant remedies in two communities. In the markets of Trujillo and Chiclayo, Evans, Téllez and Vega (2014) conducted a traceability survey of medicinal plants demonstrating that vendors have little knowledge of the source and identification of medicinal plants. Schjellerup (1991, 2009) has broadened our knowledge of late colonial ethnobotany with her work on the collections of Bishop Martinez Compagñón.

During the 1970s the World Health Organization (WHO) was very proactive in advocating the integration of Traditional Medicine into public health programs in Third World countries (WHO 1977). This culminated in the Alma Ata Declaration of 1978, which proclaimed "health for all in the year 2000" (WHO 1978; Farnsworth et al. 1985). Cabieses (2000) described his struggles to implement the UN tenets in Peru, together with Carlos Alberto Seguin (1979, 1982, 1988) who advocated the incorporation of traditional folk psychotherapy into the modern institutional framework. In 1979, they organized the First World Congress of Traditional Medicine to build on the Alma Ata Declaration. As a result of coming up with such a "hair-brained" (*descabellada*) notion, they were nearly expelled from the prestigious Colegio Médico del Perú. In addition, Peru's Minister of Public Health declined the invitation to participate in the inaugural ceremonies. In spite of these setbacks, the congress was a resounding success with participants from 23 countries and sessions in Lima, Iquitos, and Cuzco. Few medical doctors attended, however.

Building on the success of the first conference, in 1988, Dr. Cabieses presided over the Second World Congress on Traditional Medicine, which attracted 4,000 participants from 41 countries. An indication that attitudes were beginning to change in the Peruvian medical community was the fact that the Minister of Public Health, the Dean of the Colegio Médico and the Mayor of Lima all participated in the inauguration ceremony, along with a long list of university authorities. Published acts of the congress included important contributions on the medicinal flora of Peru (Pallardel 1988; Rumiche and Valderrama 1988) and for the Southern Andes (Roersch 1988). Subsequent publications of note included the southern highlands of Peru (Franquemont, Plowman et al. 1990; Roersch 1994) and the Peruvian Amazon (Duke 1994; Rutter 1994).

An important outcome of the second congress was a proposal put forth to create an Institute for the study of Traditional Medicine within the Peruvian Ministry of Public Health. This resulted in the inauguration of the Institute of Traditional Medicine (INMETRA) in 1991 with Dr. Cabieses as its first director for the next decade.

While he was director of INMETRA, Dr. Cabieses was instrumental in coordinating a network of 16 ethnobotanical gardens in Peru, which included the cultivation of medicinal plants used by traditional herbalists. He also facilitated scientific research on Traditional Medicine building a large database of herbals and monographs on 200 species of medicinal plants as well as inaugurating a national midwife program and organizing a number of international conferences. In 2001, a new administration discontinued these innovative programs.

In his last years, from his position as Rector of the Universidad Científica del Sur, Cabieses (2007) published his magnum opus on medicine in ancient Peru. He was also a strong critic of Peru's apathy regarding protection of its biocultural resources. In his book *Hoy y Ayer: Las Plantas Medicinales* (2003), he reviewed the lamentable history of medicinal plant legislation in Peru throughout the 1990s. He pointed out that the nation followed the recommendations of the US Food & Drug Administration (US FDA), which he saw as totally inapplicable, a situation traceable to the "bicultural" nature of Peruvian society where the official scientific world view predominates over traditional "cosmovision." This was occurring in spite of the fact that, since the 1970s, WHO had repeatedly formulated and refined guidelines for appropriate protection and sustainable development of medicinal plant resources and associated knowledge. Most of these recommendations were systematically ignored by the Peruvian Government. Bringing a personal perspective to bear on this matter, Cabieses (p. 118) quoted a Peruvian Minister of Public Health who stated that medicinal plants and Traditional Medicine "aren't worth a thing," and that their study was "a waste of money and effort." He ended his book (p.120) by contrasting the renewed European interest in medicinal plants with the Peruvian attitude:

> But here in Peru it's different. The lack of information and efficient research, education, and medical practice regarding the use of medicinal plants aggravates the fact that more than nine million human beings, a third of our population, in effect have as their only medical resource...the vegetal resources that surround them. The great unknown in our public health system is why so many physicians go to such lengths to exclude from their therapeutic activity the only resource that can control the suffering—not to mention the ailments—of such an important sector of our population.

During the last years that Fernando Cabieses was director of INMETRA, there emerged another initiative with regard to a rapprochement between aspects of Traditional Medicine and modern medicine. In 1999, the Peruvian social security system (EsSalud), with support from WHO and PAHO, developed the National Program of Complementary Medicine (PRONAMEN). Three Centers for Attention in Complementary Medicine (CAMECs) located in Peru's principal urban areas (Lima, Arequipa, and Trujillo) were opened. By 2012, there were 26 centers and 25 minor medical units in the country. Earlier we noted that, in 2000, PRONAMEN conducted a study demonstrating that complementary medicine had fewer side effects and better results than allopathic medicine as well as being more cost-effective (EsSalud/WHO 2000).

The earliest treatment modality of 12—inspired by Chinese traditional medicine—was phytotherapy, which is described in the Manual de Fitoterapia (2001) compiled by Dr. Martha Villar - the first and current director of PRONAMEN—together with Dr. Oscar Villavicencio under the auspices of WHO. It contains a survey of the scientific literature on the botany, phytochemistry, and pharmacology of 76 medicinal plants. It also delineates the ailments treatable by vegetal species as well as quality control and sustainable bio-agricultural production techniques. In 2009, a pilot program prescribing 20 of the 76 medicinal plants studied was initiated in "natural pharmacies" located in CAMECs in the three initial urban areas served. EsSalud has also established the Institute of Traditional Medicine (IMET) in Iquitos where modern researchers collaborate with native curanderos in investigating the properties of medicinal plants. And, in 2010, Dr. Villar collaborated with Peru's Colegio Médico in organizing the VII Congreso Nacional de Medicina Tradicional/II Congreso Mundial de Medicina Natural/ I Congreso Andino de Medicina Tradicional, Natural y Alternativa, appropriately titled "Dr. Fernando Cabieses Molina."

Issues in Ethnobotany

Moran, King and Carlson (2001) trace the emergence of biodiversity prospecting. On 5 June 1992, in order to alleviate the loss of earth's flora and fauna the Convention on Biological Diversity (CBD) was inaugurated at the UN Earth Summit in Rio de Janeiro, Brazil. CBD objectives are: 1) conservation of biodiversity, 2) sustainable use of components of biodiversity, and 3) equitable sharing of benefits derived from commercial use of genetic resources.

For biodiversity-rich developing countries the most critical element in the CBD is sovereignty over bioresources by nation states, since the treaty recognizes their right to regulate and charge outsiders for access to their biodiversity. The sovereignty component is meant to replace the "common heritage" paradigm, which provides unrestricted access to biological resources. Ideally this paradigm shift is supposed to balance the way in which all involved interest groups can gain from biodiversity use by recognizing the economic, sociocultural, and environmental values of bioresources and the cost of their preservation.

In the time since the CBD was initiated, few of the 178 signatory nations have introduced legislation requiring benefit sharing for outside commercial access to their national bioresources, although some suggestions for implementation of the CBD have been brought forward (Iwu 1996; Buitron 1999). Despite the lukewarm response to the CBD by nation states, the global shift in awareness concerning tropical deforestation provided an opportunity for ethnobotanists to assert that everyone has an interest in preserving rainforests because they might contain compounds that could cure cancer, HIV-AIDS, and other diseases (Schultes and Raffauf 1990; Elisabetsky and Castilhos 1990; Cox and Balick 1993; Schultes 1994; Brown 2003). In addition, income derived from the marketing of traditional medicinal knowledge was seen as an instrument to alleviate poverty and to finance conservation efforts (Reid 1993; Mooney 1993; Baker et al. 1995). Within a few years, however, for its critics, ethnobotany - initially seen as an instrument that could help to salvage declining traditional knowledge and biodiversity - had simply become an instrument of theft and "biopiracy."

In his book *Who Owns Native Culture?* anthropologist Michael Brown (2003) has a chapter entitled "The Ethnobotany Blues" which documents high-profile projects launched in Africa and Latin America in the early 1990s. They were organized under the U.S. initiative known as the International Cooperative Biodiversity Groups (ICBG), administered by the Fogarty International Center for Advanced Study in Health Sciences, part of the National Institutes of Health (NIH), with additional funding from the National Science Foundation (NSF) and the U.S. Agency for International Development (USAID). Projects involved partnerships between American and host-country scientists as well as major drug companies, including Monsanto, Bristol-Myers Squibb, and American Cynamid. Brown describes ICBG-Peru's troubled relationship between the Aguaruna of the Peruvian Amazon and Washington University (St. Louis), criticizing "paternalistic interventions that leave native peoples on the margins of decision-making and profit-taking" (p. 114). In Mexico, he documents how ICBG-Maya was shut down by an indigenous healers' organization and their activist allies on the grounds that it was an effort to steal native knowledge and resources. And he traces the failure of Shaman Pharmaceuticals, a California company which folded in 1999, in trying to adapt ethnobotanical bioprospecting to the "magic-bullet" paradigm of the pharmaceutical industry.

American anthropologist Cori Hayden (2003:113-116) in her ethnography on the effect of the ICBG on Mexican ethnobotany traces the impact of the UN's Alma Ata Declaration on Mexican ethnobotany. In 1975, President Echevarría established the Institute of Medicinal Plants (IMEPLAM), inaugurating an era of official recognition of traditional medicine as a legitimate resource for scientific research that integrated science and therapy. This was part of an effort to reinvigorate the National Medical Institute (1888-1910), which attempted to develop a pharmaceutical industry in Mexico based on indigenous and popular medicine. Abigail Aguilar, retired director of the national herbarium, underscores the positive impact of the UN:

What happens is that no one studies what they have. Everyone devalues what they have, especially in countries like Mexico where we've been conquered and have had another culture imposed on us...So in the case of Mexico, there's a historic complex in which everything that smelled of plants was worth nothing. Academic medical researchers weren't very interested in that kind of resource...until they heard what the WH0 said in the 1970s. That took hold in many countries, it definitely took hold here... because IMEPLAM was already in place (Aguilar in Hayden 2003:114).

Hayden's ethnography focuses on an ICBG bioprospecting agreement begun in 1993 between the University of Arizona and its pharmaceutical partners (whose contribution was a discount on the use of their equipment!) and a team of plant researchers at Mexico's National Autonomous University (UNAM) headed by ethnobotanist Robert Bye. Under the agreement, UNAM researchers sent extracts of Mexican medicinal plants to the US in exchange for research funds and promises of a percentage of royalties 10 to 20 years in the future—should a drug result from the collaboration. Influenced by the CBD, the project was also designed to collect ethnobotanical knowledge and to direct some royalties back to source-communities. It concluded in 2003 when UNAM opted out of a second renewal.

Hayden elucidates the complex issues that emerged during the project, in particular the paradoxical effects of NIH's advocacy of benefit-sharing according to the neo-liberal paradigm of bioprospecting. For NIH, this meant that field researchers were supposed to sign contracts with each individual supplier of plants. Suppliers—and, by implication, their communities—were presumed to be "authors" and "stewards" of resources as well as future benefit-recipients.

For UNAM ethnobotanists - drawing on a well-established 100-year-old research methodology and 500 years of hybridization which produced the syncretized Mestizo "herbolaria mexicana" (Mexican herbalism) - this meant collecting initial plant species from urban marketplaces and rural roadsides, a major disruption of a fundamental biopropecting assumption that plants and knowledge "come with" clearly identified local stewards, authors, and claimants. Contrasting with this vision, Bye's team used a combination of market studies, a review of the voluminous Mexican ethnobotanical literature, the 5-volume *Atlas de Plantas Medicinales* published by the National Indigenous Institute (INI), and their own historical and ethnobotanical studies to search for potentially "active" plants.

In stark contrast with the ICBG approach, there is the Mexican Institute for Social Security (IMSS) model put into practice at its Southern Center for Biomedical Research (CIBIS) in Cuernavaca and focused on the production of herbal medicines. On 20 February 1997, Hayden (2003:115) interviewed Miguel Antinori, a prominent CIBIS official who denigrated bioprospecting agreements for using Mexico's chemists as "cheap labor" and for sending extracts abroad for "more sophisticated" work. Further, he added, "It's hard to see an assertion of [Mexican] national identity in these contexts—up north, they just see Mexico as a source of raw material and certainly not as research partners or collaborators. Why don't they locate more of the development part here? Because they don't trust Mexican science."

Former Shaman Pharmaceuticals scientists Moran, King and Carlson (2001) discuss the irony in the situations described above, indicating that the majority of the biotech industry is not involved in bioprospecting, since most companies favor the use of cheaper and faster synthetic technologies over the complex process involved in exploring for natural products. Nonetheless, biotechnology spawns ethical, social, and legal debates at the margins of pharmaceutical bioprospecting, including the collaboration between big business and big science, the ethics of genetic engineering, and the patentability of life forms as well as ideas about genetics and racism, culture and ethnicity. However, it is significant to note that, since the inauguration of the CBD, no pharmaceutical bioprospecting product developed by using traditional knowledge has generated an economic profit. (But this does not mean that pharmaceutical companies do not try to impede or co-opt efforts to get natural plant products to market.) Also, only a small number of bioprospecting research expeditions begin by using ethnobotany as a discovery methodology, with the work soon evolving into economic botany as the laboratory focus shifts to the plant's chemistry, biological activity, and pharmacology/toxicology. During drug discovery, active chemical components are isolated, often modified, and patented. Patented information then becomes a commodity in itself.

Peruvian pharmaceutical researcher Pedro Angulo (2009) discusses new approaches to research on medicinal plants contrasting Western and Eastern methodologies. For example, whereas the West does not value popular wisdom and usage developed over centuries by local cultures, the East uses this knowledge as a paradigmatic base for its model of science. Whereas the West has exclusively followed the Cartesian model of scientific skepticism, Eastern pragmatism, building on tradition, has formalized usage and then applied the methods of modern science. Whereas the West has ignored traditional knowledge in designing artificial studies that isolate chemical components and evaluate their toxicity and bioactivity to later take finished products into clinical settings, the East has followed an inverse strategy, i.e., valuing traditional knowledge by applying original remedies and therapies in the medical clinic and then subjecting those that work to biochemical research and development. Whereas the West followed a basic research paradigm of random screening, component analysis and synthesis, the East recognized the holistic action of herbal medicines in seeking ways to industrialize them. As a result of the foregoing factors, Western science has developed economic botany, which uses a methodology of chemical taxonomy based on the assumption that only by knowing the chemistry of plants can we discover their active principles and bioactivity. This has led to the current emphasis on synthetic chemistry for the development of modern medicines.

Angulo (p. 363) points out that, by uncritically following the Western model for biochemical research promoted by large European and American pharmaceutical corporations,

Peru has acquiesced in the notion that countries like Peru and Mexico lack the technical and economic resources necessary to compete with foreign consortiums. As a result, these countries, for the most part, have denigrated their own indigenous knowledge and neglected the development of viable national research programs in ethnobotany and ethnopharmacology. Joining Eliane Elisabetski (1988), Angulo suggests that:

> Traditional Medicine should be the basis for the development of drugs, given that it includes the knowledge of the therapeutic value of local flora. Thus, knowledge of the practices of Traditional Medicine plays a crucial role in the selection of species to subsequently be considered as potential sources of universally applicable drugs. Elizabetsky concludes that the interaction between anthropology and ethnopharmacology is the basis on which should be developed the *holistic* investigation of medicinal plants in particular and healthcare in general.

We would only add that applied research on natural plant remedies should also be on the national agendas of Peru and neighboring republics.

Manek and Lettington in *Cultural Survival Quarterly* (2001) point out that by focusing on indigenous knowledge as it relates to the environment, the Convention on Biological Diversity managed to sidestep some of the more politically charged aspects of the intellectual property rights (IPR) issue. The greatest impact on concerns over indigenous and local-community rights can be traced to the mercurial rise of biotechnology on the international trade front and the 1995 version of the World Trade Organization (WTO) Agreement on Trade Related Aspects of Intellectual Property Rights (TRIPS). These two factors have created a large potential market for indigenous and local knowledge and resources, while at the same time raising concerns about the risk that these resources will be misappropriated. Thus this knowledge is receiving increasing international attention in terms of its relationship to human rights as well as its relevance to modern science. The situation has created opposing pressures calling for the rights of local and indigenous peoples on the one hand and further exploitation of their knowledge on the other. Moran, King and Carlson (2001), Manek and Lettington (2001), Greaves (1995), and Brush and Stabinsky (1996), indicate that the biggest problem with the orthodox intellectual property system is its focus on material aspects of knowledge at the expense of the cultural. They advocate recognition of alternative worldviews in the formulation of new indigenous knowledge rights that are localized, relevant, pertinent, and effective.

In their article in *Cultural Survival Quarterly* Bannister and Barrett (2001) contend that bioprospecting is a form of economic botany that can run contrary to the ethnobotanical objectives of protecting biological and cultural diversity. The economic focus of this activity highlights issues concerning indigenous rights, cultural knowledge, and traditional resources - areas in which current intellectual property protection regimes are inadequate and inappropriate. However, indigenous communities are increasingly forced to employ intellectual property rights to protect these resources. Protection issues ought to be addressed well before the point at which employing intellectual property mechanisms seems to be the only alternative. Significant control lies at the point of decision about publication and dissemination of knowledge to the wider community, which raises important questions about facilitating the appropriation of cultural knowledge. The authors (p. 10) advocate a more "precautionary" approach to ethnobotanical inquiry in assisting indigenous communities in protecting their cultural heritage and intellectual property rights.

Probably the major concern in many traditional communities is that their spiritual legacies will be profaned by a secularized and consumer-driven outside world. Often, however, legitimate economic considerations also play a role in the defensive reactions of these societies to the well-intended but naïve desire of the academic world to place its findings in the public domain. Brush and Stabinsky (1996) and others (Greaves 1995; Bannister and Barrett 2001) have warned that the downside in this approach is that a "colonializing archive" can become easily "mined" for clues in the search for new drugs without the inconvenience of fieldwork or inclusion of source communities in the benefits derived from products resulting from research.

However, although acknowledging genuine concerns about neocolonialism and biopiracy, we would submit that each situation has to be considered on its own merits, especially with regard to its specific cultural context. A first step in the evaluation process should involve the important distinction between "indigenous peoples" and "local communities" (Moran, King and Carlson 2001). The latter for the most part are farmers who speak the national language, practice the majority religion, and identify with the nation-state, especially with regard to their socioeconomic aspirations, whereas the former tend to be tribal and/or ethnic minorities, who seek collective rights and self-determination for their biological and cultural resources. Although it is often the case that in both communities traditional knowledge and resources are undocumented and in danger of disappearing, this danger tends to be more pressing in local communities as their members continue to adapt to privatization and globalization. In cases such as these successful ethnobotanical intervention requires a methodology that combines "salvage ethnography" with "rapid assessment". This is the methodology that we initially applied in Peru, motivated by our prior experience in Southern Ecuador where traditional knowledge of medicinal plants similar to those found in Northern Peru is diminishing at an alarming rate. However, with our database firmly established as a research vehicle, we can now turn our attention to facilitating proactive issues of education, conservation, and sustainable development of natural plant products.

India provides a positive example of the proactive application of this approach. Taking advantage of the "novelty" criterion in international patent law, with regard to the documentation of Ayurvedic and other traditional medicine, millennial Sanskrit texts as well as modern publications are included in a traditional knowledge database, which is subsequently provided to patent agencies. The expectation is that, by placing the knowledge about long-term cultural precedents for traditional uses in the public domain, this research will prove that contemporary patent applications derived from local medicinal knowledge lack originality, i.e., that they are not "novel" enough to qualify as inventions warranting protection under international patent law, and are thus not patentable.

Fortunately, in 2002, with the support of the International Phyto-Genetic Resource Institute (Rome, Italy), Peru promulgated Law 27811 for the protection of the collective knowledge of indigenous peoples related to biological resources. Article 17 of the law establishes a National Public Register to include collective knowledge that is in the public domain. This register is administered by INDECOPI (National Institute of Competitive Defense and Intellectual Property), with the obligation to send the information recorded to principal patent offices around the world, a protective defense mechanism intended to prevent the granting of patents which do not meet the criteria of novelty and degree of inventiveness (Venero 2005b: 32-36).

In the first decade of the 2000s—although little had been done to protect and sustainably develop these valuable natural resources and associated knowledge—increasingly unfettered access was being granted to foreign pharmaceutical corporations. In 2004, a forum organized by the Peruvian Congress and the Sociedad Peruana de Derecho Ambiental (SPDA), an NGO dealing with environmental law, pointed out that foreign patent applications were pending or granted for 19 Peruvian plants, and that the Government was not making resources available to determine if the patents or claims met the requirements of patent law (Venero 2005a: 54). Adding insult to injury, on 28 March 2009, *Somos*, the news magazine of the prestigious daily *El Comercio* reported that, under the terms of the Peruvian-North American Free Trade Agreement, claims by American pharmaceutical companies were to be granted "exclusive protection" for alleged "new products" regardless of whether or not they qualified or had prior licenses or patents (Chumpitas 2009).

A classic example of one hand not knowing what the other is doing was revealed on 16 July 2009 when Zoraida Portillo of SciDev.Net reported that Peru had denied patents to companies from France, Japan, South Korea, and the US on the grounds that their products were developed using traditional knowledge. The denials emanated from the Peruvian National Commission Against Biopiracy discussed in the Peruvian Congressional Forum of 2004 and created under Law 28216. However, the Portillo article ended with a quote from Michel Pimbert of the International Institute for Environment and Development: "It would be naïve to think that national governments would automatically share benefits with local communities when biopiracy is prevented or compensation obtained."

In the Congressional Forum of 2004 which discussed the formation of the Commission a number of important issues were addressed, including intellectual property (Bazán 2005: 21-35), the high protein cereal Quinua and biopiracy (Caillaux 2005: 36-47), passage of the law creating the commission for the protection of access to Peruvian biodiversity and the collective knowledge of indigenous peoples (Morales 2005: 48-49), and efforts to annul the US patent for the virility stimulant Maca (Venero 2005a: 50-55) as well as suggestions for combating biopiracy (Venero 2005a: 74-78). Briefly noted was the issue of genetically modified foods (Agurto 2005: 71-72), anticipated as a concern that was likely to emerge with approval of a free trade agreement with the US. When the Commission was legally mandated, 19 plant claims were slated for review. By 2010, claims for 69 plants were being researched, 17 cases of biopiracy had been identified, and seven had been successfully blocked (Smallwood 2011: 36-37). One hopes that in all these deliberations the following remarks by forum panelist Jorge Agurto (2005: 71) will be borne in mind:

The problem underlying biopiracy is the open recognition of the rights of the indigenous peoples and communities. Many times they have been excluded and marginalized from the politics of Government. Even today we encounter members of Congress who are either unaware of the existence of indigenous peoples or who do not recognize their rights. It is impossible to speak of biopiracy if we do not defend the holders of many genetic resources, those who have achieved the domestication, knowledge, and technology to utilize biodiversity in a sustainable fashion. They are also the holders of the right to prior informed consent, a fundamental right to know the objectives of the exploration and exploitation of their resources and traditional knowledge and the consequences or potential benefits that can come with industrial, commercial or scientific uses.

Spanish anthropologist Luisa Abad (2003: 274) concludes in her book *Ethnocide and Resistance in the Peruvian Amazon* that foreign and domestic development policies contribute to the marginalization of indigenous peoples:

Underdeveloped, developing, Third World, North-South..., perhaps the language has been changing in these times and the terminology has been adapting itself to partially new habits, but the unequal, hierarchical reality remains the same, given that those who exercise power continue to be the same. International assistance also keeps promoting unequal development between peoples.

Biodiversity Conservation and Traditional Medicine

A policy report, *Biodiversity, Traditional Knowledge and Community Health: Strengthening Linkages,* published by the United Nations University - Institute of Advanced Studies in Yokohama, Japan addresses many of the issues discussed above (Unnikrishnan and Suneetha 2012). Building on the WHO Alma Ata Declaration of 1978 relating to Traditional Medicine and primary healthcare, the UN Convention on Biological Diversity of 1992, and the UN's Middle Development Goals (MDGs) of 2011, this document shows that links between Traditional Medicine and biodiversity are strengthened by three processes: 1) a medical approach involving national efforts to integrate Traditional Medicine into institutional healthcare delivery which includes challenges related to safety, quality, efficacy, access, and regulation; 2) a market-oriented approach focused on drug development or tourism promotion focused on biomedical products and services as marketable commodities; and 3) a community-focused approach activated by civil society organizations focused on conservation implemented through a grassroots mobilization process involving health professionals, botanists, conservationists, and community activists.

The community-based approach shows allegiance to the Alma Ata primary health care model. Examples include the barefoot doctors strategy in China and the social health activist programs in India. Given the centrality of biodiversity in human lives, there still is a need to develop sustainable strategies for health maintenance combined with conservation of biological resources and linked to local knowledge and practices. This is relevant even in developed countries where there is an increasing demand for alternative and complementary medicine.

At the beginning of the UNU report in a "Message from the Director," Govindan Parayil (p. 6) assesses progress towards the CBD agenda of a global development path that is sustainable, equitable, environmentally just, and economically rewarding. He sees that the prognosis is not encouraging. Progress has been made, but we still are falling far short in even sustaining current levels of well-being. "Negative environmental trends continue to be exacerbated by human interventions—primarily led by a model of unsustainable and conspicuous consumption." He adds: "The extraordinary emphasis on developing produced capital appears to have overwhelmed all other aspects of natural capital required for our well-being."

On the positive side, Parayil notes increased awareness of the gap between planning and implementation. Welcome signs of change include "increasing resolve to align production activities with environmental and equity considerations" as well as "efforts aimed at reforming global institutional structures to create more synergies and effective implementation of relevant policies." He concludes:

Current accepted standards of practice and business norms must be re-oriented to include a more consultative policy setting with all major actor representatives. [This] would require designing regulations that acknowledge the need for balance among all forms of capital, and incentives that provide equitable access to resources and services.

The UNU policy report documents 30 successful community-based projects from around the world. Despite their success in finding workable solutions to meet conservation and primary health care needs, the scale of operation of these programs has not been enhanced or expanded. This is due to a number of factors listed in the report, some of which include:

- *There is a clear need to include ecological, conservation, and sociocultural factors in goal-setting related to health and development programs.*

- *High external dependency, especially in pharmaceuticals and medical technologies, disincentivizes local innovations in Traditional Medicine and healthcare.*

- *Through a top-down health care approach, societies have organized themselves to be more disease-centric (with supporting institutions, research, industry, government departments, strategies, and programs) than wellness-centric. A paradigm shift in the mind set as well as in systems and structures to wellness (prevention/promotion) is a challenge, yet essential. For this to occur, internalization (not mere awareness) and implementation are essential.*

- *Traditional health promotion and related conservation schemes focused chiefly on medicinal plants [have been] seen more as avenues for economic development and hence expected to become self-supporting ... To realize self-sufficiency, costs of delivery of various related services from resource collection to distribution and infrastructure to identify and support healers need to be thought out comprehensively. There is a critical need for innovative approaches for funding mechanisms in this area.*

- *At the policy level, there appears to be a tendency towards non-realistic target setting. Implementation mechanisms for such targets still rely primarily on formal mainstream processes such as modern infrastructure and trained doctors, while including and appropriately training specialists outside the "modern" realm of training, especially at the community health worker level, might have hastened the processes to achieve various goals. A reflexive learning approach to development is especially important in this context where no single knowledge system has sufficient conceptual, theoretical or practical authority in addressing health challenges.*

- *While attempts to document and protect traditional medical knowledge in searchable and other inventories are important in terms of defensively protecting such knowledge from misappropriation, efforts to use such knowledge to augment community health are still insufficient. Attempts to open such inventories for research purposes still play into mainstream drug development processes—more than local healthcare. This stymies efforts to use and expand such initiatives to provide better community and public healthcare.*

- *High erosion of traditional knowledge and lack of perceived support for traditional health practitioners have evidently led to a decrease in the receptivity to and transfer of all aspects of such knowledge between generations. It has been observed that in cases where the knowledge system is perceived to bring in recognition and supplemental income, younger generations are keen to learn, develop, and sustain them.*

- *We often see that the dominant education and research systems tend to enhance knowledge and technologies using universal standards, without much attention to the capacities and needs of specific regions or populations ...* [*resulting in*] *a dearth of comprehensive theoretical approaches to assessing . . . traditional knowledge* [*which*] *is believed to be key to the disregard of traditional cultures ... This then calls for the design and implementation of culturally appropriate pedagogical methods with an intercultural inclination and trans-disciplinary approach and their integration into formal and informal learning processes.*

- *There appears to be a clear need for designing a radical and innovative approach to integrate Traditional Medicine into mainstream health systems. This would further require full institutional backing from various related governmental and non-governmental agencies that link supply chains of medicinal resources with health practitioners and consumers with the highest standards of quality, safety, and efficacy.*

With regard to a plan of action, this policy paper advocates the use of integrated rapid assessment protocols similar to those used in some of the case studies outlined in the report - duly adapted to local cultural and environmental circumstances. It provides an assessment framework and the following "potential strategies":

- Assessment methods to inventory resources and knowledge used in healthcare.
- Knowledge validation, generation, and use.
- Capacity building for different stakeholders.
- Cross-learning between different knowledge systems.
- Mechanisms to protect traditional resources and knowledge.
- Linkages with economic development objectives.
- Expansion of partnerships with different stakeholders.
- Effective communication strategies.
- Synergizing community initiatives with civil society organizations and policy processes in identifying critical areas of engagement.

Complementing the positive examples from the UN University-Yokahama report are the lessons learned from a failed project in Northern India which sought to develop a medicinal plant value chain between local Himalayan farmers and a Dutch company (Ayurveda Health) in a project undertaken by The Royal Tropical Institute (KIT) and the Center for Sustainable Development (CSD) of the Netherlands in cooperation with local government agencies (Alam and Belt 2009). The authors point out that, worldwide, medicinal plants are being depleted at a rapid pace due to large-scale, unsustainable collection from natural habitats. Conservation of these species is critical for four reasons: 1) they are a source of natural ingredients used by the manufacturers of modern pharmaceuticals resulting in a large and increasing demand (Balick and Mendelsohn 1992; Lambert et al. 1997; FAO 1997); 2) medicinal plants form the basis of homeopathy and traditional medicines, and, along with traditional knowledge, are crucial for traditional healers, who play a vital role in the lives of poor people and their animals in developing countries (WHO 1999; Hamilton 2008); 3) the collection and marketing of medicinal plants is a valuable source of livelihood for large numbers of poor people in developing countries; and 4) medicinal plants are an essential component of biological diversity and conservation (SCBD 2001).

Regarding lessons learned, three reasons are given for the project's lack of success: 1) poor quality planting material supplied to farmers resulting in high mortality of plants; 2) too many uncoordinated farmers planting uneconomic plots on marginal land which resulted in low upkeep motivation and unrealistic expectations that were not realized; and 3) poor understanding of local farming dynamics and the emergence of a successful alternative cash crop. These are factors that should be evaluated in any efforts to build a successful supply chain for CAMEC-EsSalud Trujillo.

Work to Date

Work up to 2012—besides developing a database of 510 medicinal plants (Bussmann and Sharon 2006b, 2007b, 2009c) and 974 remedies of mixtures (Bussmann, Glenn, Meyer, Kuhlman, and Townesmith 2010)—has demonstrated that herbal commerce in Peru is a major economic resource (Bussmann, Sharon, Vandebroek, Jones and Revene 2007), which, although used alongside modern pharmaceutical products, is showing signs of diminished popular knowledge of applications (Bussmann, Sharon, and Lopez 2007; Bussmann, Sharon, and Garcia 2009). Laboratory research on most of the database has ranged from minimum inhibition concentrations (Bussmann, Malca et al. 2010) to toxicity screening (Bussmann, Malca et al. 2011) as well as bioassays to determine antibacterial activity (Bussmann, Sharon et al. 2008; Bussmann, Glenn et al 2009a; Bussmann, Glenn et al. 2010) and phytochemical analysis (Bussmann, Glenn et al. 2009b; Perez, Rodriguez et. al. 2012) with more focused analyses of herbal treatments for acne (Bussmann, Sharon et al. 2008), malaria (Bussmann and Glenn 2010), and kidney problems and urinary infections (Bussmann and Glenn 2011). Other studies have sought to identify Ulluchu, a ceremonial plant of the pre-Hispanic Moche culture (Bussmann and Sharon 2009b) as well as surveying colonial sources of medicinal plants in Northern Peru and Southern Ecuador (Bussmann and Sharon 2009a). An ethnography of peasant herbalists which documented aspects of the market supply chain showed that suppliers are not adequately remunerated and revealed threats posed by lack of conservation measures and overharvesting (Revene, Bussmann and Sharon 2008). Carrillo (2012) criticized the scientific reductionism of laboratory research in attempting to appropriately verify traditional remedies. Smallwood (2010) conducted an ethnography of the work of the Peruvian National Commission Against Biopiracy. Anthropological studies of traditional *curanderos* and their curing altars *(mesas)* included articles by Sharon (2009); Sharon and Gálvez (2009); Sharon, Glass-Coffin and Bussmann (2009); and Glass-Coffin, Sharon and Uceda (2004).

It is worth noting that, during the decade that we have been working in the field and the laboratory, there has been a sea change in attitudes and perceptions of Traditional Medicine (Sánchez Garrafas, eds., 2009; Vergara and Vásquez, eds., 2009). As noted earlier, in Trujillo, Lima and Arequipa, a pilot program prescribing medicinal plants scientifically validated by WHO/PAHO has been initiated by EsSalud's National Program for Complementary Medicine, an initiative begun in 1999 with three centers which has grown to 26 to date (Fernández 2009; Villar and Villavicencio 2001). In Trujillo, the Missouri Botanical Garden (MOBOT) Sacred Seeds program started an herbal garden and educational outreach program at the site museum of the pre-Hispanic Chimú city of Chan Chan. University of Trujillo (UNT), botanists Mostacero, Castillo, Mejía, Gamarra, Charcape and Ramírez (2011) compiled a volume of 774 medicinal plants, including taxonomy, ecogeography, fenology, and ethnobotany. In Huamachuco, a program of ethnobotany and conservation manifest in community gardens and seed banks of medicinal and food plants is slowly emerging through collaboration between a local peasant community, the Beneficencia Publica, the regional hospital, EsSalud, MOBOT's Sacred Seeds program, MHIRT, and the Peace Corps. Future work will involve developing a supply chain between Huamachuco and EsSalud's Centro de Atención de Medicina Complementaria in Trujillo with scientific validation by MOBOT, University at Buffalo (SUNY), and the Biotransformation and Natural Products Laboratory at UNT.

Plant Nomenclature in Northern Peru

The naming of plant species follows three general patterns. Plant names already used by original indigenous populations are often maintained, although slightly modified. Plants similar to species already known, or with similar habitus, often receive the same name (transposition). In other cases, completely new names are created (neology) (Van den Eynden et al. 2004).

The vernacular names of the plants used in Northern Peru reflect the historical development of plant use in the region. Introduced species (e.g., *Apium graveolens* - Apio, *Foeniculum vulgare* - Hinojo), native species similar to species found in Spain (e.g., *Adiantum concinnum* - Culantrillo, *Matricaria frigidum* - Manzanilla), as well as species growing mostly in the coastal regions of the area (e.g., *Alternanthera porrigens* - Sanguinaria), are often addressed with names derived from Spanish roots. Plants from the mountain forests and especially the Andean highlands or the Amazon are often known by their Quechua names (e.g., *Pellaea ternifolia* - Cuti Cuti, *Amaranthus caudatus* - Quihuicha, *Banisteriopsis caapi* - Ayahuasca), and a few plant names can be traced back to Mochica (the original indigenous language spoken on the coast of Northern Peru) (e.g., *Nectandra* spp. - Espingo) (Bussmann and Sharon 2009c). Van den Eynden et al. (2004) observed similar patterns in Southern Ecuador, although her study focused only on edible species. Nine hundred thirty-eight vernacular names were recorded for 510 plant species. About one third of all names represented Quechua names or had Mochica roots, while 66.5% of all names were of Spanish origin or at least had Spanish components. In comparison, 41% of the vernacular names of edible plants in Southern Ecuador were found to be of Spanish origin. More than half of the indigenous species had only one vernacular name, with the remaining species hading a variety of indigenous names, often derived from the same root. In comparison, almost 75% of the introductions were known by one name only. The slight differences in plant names indicate that the species have been used in the region for a long time and that their names reflect small variations in the local dialects.

Two Decades of Ethnobotany in Northern Peru and Southern Ecuador

Since 1995 ethnobotanical data were collected in Northern Peru and Southern Ecuador (Fig. 1), from plant vendors while purchasing plant materials in local markets, by accompanying local healers to the markets when they purchased plants for curing sessions, and in the field when they were harvesting. In addition, plants were collected by the project members in the field, and - together with the material purchased in the markets - taken to the homes of *curanderos* to discuss healing properties, applications, harvesting methodology and origins. The project region represents a floristic hotspot in Peru, with striking gradients from coastal deserts and dry forests, to some of the wettest Paramos known as well as to the Amazon lowlands (Fig. 2). At the *curanderos'* homes the authors also observed the preparation of remedies and participated in healing rituals. Plant uses were discussed in detail with informants, after seeking prior informed consent from each respondent. Following a semi-structured interview technique, respondents were asked to provide detailed information about the vernacular plant name in Spanish or Quechua, plant properties (hot/cold), harvesting region, ailments for which a plant was used, best harvesting time and season, plant parts used and mode of preparation/applicatio, including the addition of other plants. All interviews were carried out in Spanish, with at least one of the authors present. Both authors are fluent in Spanish and no interpreter was needed to conduct the interviews. Data on plant species, families, and additional vernacular names were also recorded.

Many of the species reported from Northern Peru are widely known by *curanderos* and herb vendors as well as the general population of the region and are employed for a large number of medical conditions. One hundred fifty to two hundred plant species, including most of the introductions, are commonly sold in the local markets (Bussmann and Sharon 2006b). Rare indigenous species were either collected by the healers themselves, or were ordered from specialized collectors or herb vendors. The same plants were frequently used by a variety of healers for the same purposes with only slight variations in recipes. However, different healers might give preference to different species for the treatment of the same medical condition. All species found were well known to the healers and herb vendors involved in the study, even if they themselves did not use or sell the species in question. Many species were often easily recognized by their vernacular names by other members of the population. This indicates that these remedies have been in use for a long time by many people. The use of some species, most prominently San Pedro (*Echinopsis pachanoi*), Maichil (*Thevetia peruviana*) and Ishpingo (various species of *Nectandra*), can be traced back to the Moche culture (AD 100-800). Representations of these plants are frequently found on Moche ceramics, and the remains of some were found in a variety of burials of high-ranking individuals of the Moche elite, e.g., the tomb of the Lord of Sipán (Bussmann and Sharon 2009c).

Medicinal Uses

Five hundred ten plants with medicinal properties were registered in Northern Peru. The same species was often used for various medical conditions and applied in different ways for the same condition. For example, nervous disorders might be treated using different parts

Figure 1: Study area: Peruvian Departments of Amazonas, Piura, Lambayeque, La Libertad, Cajamarca, San Martin and the Ecuadorian Province of Loja. Dry forest (above) and Cloudforest (below) indicating the striking moisture gradient.

Figure 2: A 50km-transect in Southern Ecuador from dry forest (250mm of precipitation, left) to Podocarpus National Park Paramo (10000mm precipitation, middle) to Amazon lowlands (right).

of a plant in different applications, e.g., topical (as a Poultice or Bath), oral (ingestion of plant extracts) and by supplying the patient with a *seguro*, a bottle filled with herbs and perfumes which serves as a protective charm. Two thousand four hundred ninety-nine different uses were registered for the 510 species encountered. In the following, the total number of uses/applications and the number of species used are given, rather than only the number of plant species used to treat a condition, in order to emphasize the importance of the treatment of specific conditions.

The highest number of species (207, 40.4%) is used for the treatment of "magical" ailments, with 682 (27.3%) of all conditions. Respiratory problems (91 species, 18.5%) were mentioned as 233 (9.3%) of all uses; 98 species (19.1%) are used to treat psychosomatic and nervous system problems, with 176 applications (7%). Kidney and Urinary tract disorders are treated with 69 species (16.6%), for 111 conditions (4.4%). Rheumatic and arthritic symptoms are mentioned in 103 uses (4.1%), with 55 species (8.8%) used for treatment. Infections of female organs are treated with 105 species (20.9%) and comprised 100 (4.4%) of all conditions.

Treatments are most often performed in the homes of the individual healers, who normally have their *mesas* (healing altars) set up in their backyards (Fig. 3). Healers also treat patients at altars and consultation chambers (*consultorios*) in their homes, at sacred sites in the countryside, or at sacred lagoons high in the mountains. Healing altars bearing a large number of power objects are often employed (Fig. 4). A curing ceremony normally involves purification of the patient by orally spraying blessed herbal extracts on the whole body to fend off evil spirits and by nasal ingestion of tobacco juice and perfumes. Plant supplies come mostly from the local markets (Fig. 5). Two hundred seventy-eight different medical conditions were recorded. Most plants were used for the treatment of multiple ailments. The large variety of conditions is grouped into 72 main categories.

Magical Uses

Mental, neurological, and psychosomatic disorders are highly prevalent on a global scale. The burden of mental health problems has been seriously underestimated. Although neurological problems are only responsible for about 1% of global deaths, they contribute to

Figure 3: Contemporary healing altar (*mesa*) in Southern Ecuador

over 11% of the global disease burden. It is estimated that this share will rise to 15% by 2020 (WHO 2009). Western medicine often offers little help for patients afflicted by these disorders.

Healing altars (*mesas*) in Northern Peru often follow the ancient traditions by including a large variety of "power objects," frequently with a "pagan" background. Objects such as seashells, pre-Columbian ceramics, staffs, stones, etc. are very common on Peruvian *mesas* and

Figure 4: Contemporary healing altar (*mesa*) in Northern Peru

Figure 5: Plant preparations and raw plants for sale in Mercado Mayorista, Trujillo.

are blended with Christian icons such as crosses and images of saints. Treatments are most often performed in the homes of the individual healers, who normally have their *mesas* set up in their backyards. Healers also treat patients at altars and consultation chambers (*consultorios*) in their homes, at sacred sites in the countryside, or at sacred lagoons high up in the mountains. A curing ceremony normally involves purification of the patient by orally spraying blessed herbal extracts on the whole body to fend off evil spirits and by "Spiritual Flowerings" (*baños de florecimiento*). In most cases, the cleansing of the patients involves drinking boiled *San Pedro* juice and the nasal ingestion of tobacco juice and perfumes. Sometimes extracts of *Brugmansia* spp., and tobacco are also used to purify the patients. While the incantations used by healers during their curing sessions include Christian components (e.g., the invocation of Christ, the Virgin Mary and a variety of saints), references to Andean cosmology (e.g., to the *apus* or the spirits of the mountains) are very common. The use of guinea pigs as diagnostic instruments is standard in Northern Peru (Sharon 1978, 1980, 1994, 2000; Joralemon and Sharon1993).

Traditional Medicine is also gaining more attention from national governments and health providers. Peru's National Program in Complementary Medicine and the Pan American Health Organization recently compared Complementary Medicine to allopathic medicine in clinics and hospitals operating within the Peruvian Social Security System (EsSalud 2000).

Mal Aire (Bad Air), *Mal Viento* (Bad Wind), *Susto* or *Espanto* (Fright), *Mal Ojo* (Evil Eye) and *Daño* or *Brujería* (Sorcery) are seen as very common illnesses in Andean society. Causes include sudden changes in body temperature (*Mal Aire, Mal Viento*), any kind of shock (*Susto, Espanto*), ¨humors¨ or spells cast by other people (*Mal Ojo*) and poisoned food, curses, etc. (*Daño, Brujeria*). Medical problems caused by outside influences were reported in a wide variety of studies (Girault 1987; Oblitas 1992). The Western concept of "psychosomatic disorders" comes closest to characterizing such illnesses.

These illness categories are deeply rooted in Andean society, and Western medicine does not offer efficient alternatives to traditional treatment. This might explain why this category still has such outstanding importance. Treatment in many cases involves the participation of the patient in a cleansing ceremony or *limpia*. This could either be a relatively simple spraying with perfumes and holy water or an all-night ceremony focused on the healer's curing altar (*mesa*). In the days after a ceremony patients are normally treated with a *baño de florecimiento* (spiritual flowering bath) in order to relieve them of any remaining negative symptoms or spirits. In addition, patients frequently receive *seguros* (herbal amulets) for protection against further evil influences and for good luck. *Seguros* are flasks or bottles filled with powerful herbs as well as perfumes, saints' images, and the hair and fingernails of the patient.

The large number of plant species used for the treatment of psychosomatic disorders indicates that the *curanderos* of Northern Peru are valued specialists who are consulted mainly for these conditions. This is all the more interesting since Western medicine has still not found efficient treatments for psychosomatic disorders. The plant species used for "magical" or "ritual" disorders come mostly from the high Andes, especially from the vicinity of sacred lakes, since plants from those regions are regarded as especially powerful. This links the present day curing practices directly to ancient Andean cosmology. The use of purgatives and laxatives to literally "expel" evil spirits is also very common.

In total, 98 plant species belonging to 172 genera and 78 families were documented and identified as herbal remedies used to treat nervous system problems in Northern Peru. Most species used were Asteraceae (36 species, 16.21%), followed by Solanaceae (15 species, 6.76%) and Lamiaceae (14 species, 6.31%). The most important families dealing with the nervous system are somewhat over-represented in comparison to the overall medicinal flora, while some other medicinally important families (e.g., Poaceae, Cucurbitaceae, Euphorbiaceae) are completely missing or under-represented from the nervous disorders category (Bussmann and Sharon 2006b).

The majority of herbal preparations were prepared from the whole plant (31.56%), while the leaves (24.48%), stems (21.24%) and flowers (8.55%) were used less frequently. Whole plants and stems were more often used than is characteristic for the overall medicinal preparations found in the region (Bussmann and Sharon 2006b). This indicates that the local healers count on a very well developed knowledge about the properties of different plant parts. In over 60% of the cases fresh plant material was used to prepare remedies, which differs slightly from the average herbal preparation mode in Northern Peru. Interestingly, only about 36% of the remedies were applied orally, while the majority were applied topically (46.65%), often as a bath, and the remaining ones were used as spiritual safeguards ("*seguros*"). This is different from the regional average of application and underscores the importance of spiritually oriented treatments. Over 79% of all remedies were prepared as mixtures obtained by boiling plant material either in water or in sugarcane alcohol.

Respiratory System

WHO reports that respiratory illnesses are high on the list of importance as causes of death and morbidity on a global scale and thus has elaborated a Strategy for Prevention and Control of Chronic Respiratory Diseases (CRDs) (WHO 2002). Respiratory problems are a major cause of infant deaths in Peru (Weil 1978).

In total, 91 plant species belonging to 82 genera and 48 families were documented and identified as respiratory system herbal remedies in Northern Peru. Most species used were Asteraceae (15 species, 16.67%), followed by Lamiaceae and Fabaceae (8.89% and 5.56%). Most other families contributed only one species each to the pharmacopoeia. The most important families are well represented in comparison to the overall medicinal flora, although some other medicinally important families (e.g., Euphorbiaceae, Lycopodiaceae, Cucurbitaceae) are completely missing from the respiratory category (Bussmann and Sharon 2006b).

The majority of respiratory disorder remedies were prepared from the leaves of plants (27.69%), while the whole plant (18.46%), flowers (13.85%) and stems (17.69%) were used less frequently. This indicates that the local healers count on a well developed knowledge of the properties of different plant parts. In almost 55% of the cases fresh plant material was used to prepare remedies, which differs little from the average herbal preparation mode in Northern Peru. About 86% of the remedies were applied orally, while the remaining preparations were applied topically. Over half of all remedies were prepared as mixtures of multiple ingredients by boiling plant material either in water or in sugarcane alcohol.

Respiratory disorders are so common globally and over-the-counter remedies, both allopathic and complementary, are so frequently sold that much effort has been put into the verification of traditional remedies. Almost 50% of the plants or their congerers found in the respiratory pharmacopoeia of Northern Peru have been studied for their medicinal properties. The original hypothesis applied in these studies was that many species employed for respiratory illnesses would be non-native and were introduced to treat diseases that were originally also introduced by colonialists. However, this hypothesis was not validated. Quite the contrary, many remedies for respiratory ailments are native to the study area. From this perspective it is surprising to see how many species have actually been studied at least in a preliminary fashion.

Urinary System (Kidneys, Bladder)

The WHO (2005) report on urinary tract infections (UTI) indicates that they are one of the most commonly occurring bacterial infections, particularly in children. It has been estimated that UTIs are diagnosed in 1% of boys and 3-8% of girls. In the first year of life UTI is more prevalent in boys with rates of 2.7% compared with 0.7% in girls. The reported rate of recurrent UTI is around 12-30%, with greater risks in infants < 6 months of age. Studies have shown a higher UTI prevalence of 8-35% in malnourished children, with the risk of bacteria increasing significantly in accord with the severity of malnutrition.

A total of 69 plant species belonging to 61 genera and 43 families were documented for kidney and urinary-tract problems in Northern Peru. Most species used were Asteraceae (8 species, 11.43%), followed by Fabaceae and Poaceae (both 5 species each, 7.14%). All other families contributed only one species each to the pharmacopoeia. The most important families are similarly represented as in the overall medicinal flora, while some other medicinally important families (e.g. Lycopodiaceae, Cucurbitaceae) are completely missing from the kidney category (Bussmann and Sharon 2006b).

The majority of kidney herbal preparations were prepared from the whole plant (27.78%), while the leaves of plants (25.56%), flowers (12.22%) and stems (16.67%) were used less frequently (Bussmann and Sharon 2006b). This indicates that the local healers count on a very well developed knowledge of the properties of different plant parts. In almost 64% of the cases fresh plant material was used to prepare remedies, which differs little from the average herbal preparation mode in Northern Peru. About 88% of the remedies were applied orally, while the remaining ones were applied topically. Over half of all remedies were prepared as mixtures of multiple ingredients by boiling plant material either in water or in sugarcane alcohol.

Kidney and urinary system problems are very common globally, but allopathic treatments, in particular with regard to renal calculi, are mostly focused on dilation of the ureter and pain management. Although a large number of plants are used in Traditional Medicine to treat this problem, less than 35% of the plants found in Peru or their congeners have been studied at all for their medicinal properties. Kidney and urinary tract diseases are a major health challenge worldwide. Many plant species are traditionally used for kidney disease treatment, and some have been investigated for their efficacy with positive results. An often-limiting factor to these investigations is lack of comprehensive ethnobotanical data to help choose plant candidates for potency/efficacy tests. Since the plant parts utilized in preparation of kidney remedies are reported in this survey, it can serve as an indication of species that may need further ecological assessment on their regeneration status.

Rheumatic Problems

The National Institutes of Health (NIH) report that an estimated 23.5 million Americans suffer from auto-immune diseases and that this number is expected to grow. Medical research has currently identified 80-100 auto-immune diseases, and 40 additional diseases are suspected to have an auto-immune basis. Auto-immune diseases collectively rank in the top 10 leading causes of death for women aged from adolescence up to age 64. In Western medicine, the most common treatments are immuno-suppressants, which are known to have devastating long-term side effects (Gillett et al. 2000).

Drafty housing conditions as well as difficult working conditions, lead to a wide spectrum of muscular-skeletal disorders, including rheumatism, arthritis, and bone and muscle pain. In total, 55 plant species belonging to 53 genera and 43 families were documented and identified as autoimmune herbal remedies in Northern Peru. Most species used were Fabaceae (4 species, 7.27%), followed by Rosaceae and Myrtaceae (both 3 species each, 5.45%). All other families contributed only one or two species each to the total. The most important families are clearly over-represented in comparison to the overall medicinal flora, while some other medicinally important families (e.g., Asteraceae, Lamiaceae, Euphorbiaceae, Apiaceae, Lycopodiaceae, Cucurbitaceae) are less commonly used for the treatment of auto-immune disorders and pain or are completely missing from the category (Bussmann and Sharon 2006b).

The majority of the herbal preparations were prepared from the leaves of plants (35%), while the whole plant (21.25%) and stems (17.5%) were used less frequently. This indicates that the local healers count on a very well developed knowledge about the properties of different plant parts. In 60% of the cases fresh plant material was used to prepare remedies, which differs little from the average herbal preparation mode in Northern Peru. Only about 55% of the remedies were applied orally, while the remaining ones were applied topically. This differs little from the regional average of application. Over half of all remedies were prepared as mixtures of multiple ingredients by boiling plant material either in water or in sugarcane alcohol.

Very little scientific evidence exists to prove the efficacy of the species employed as remedies to treat auto-immune problems. Less than 22% of the plants found or their congeners have been studied for their medicinal properties.

Internal Organs (Liver, Gallbladder)

Disorders of internal organs fall far behind as the most commonly treated medical conditions (Bussmann and Sharon 2006b). This is an indication that *curanderos* in Northern Peru are to a large extent specializing in the treatment of psychosomatic disorders, and that "bodily" illnesses are treated more as a sideline. However, a large number of plant species were used by local healers to treat liver and Gallbladder ailments.

In total, 51 plant species belonging to 43 genera and 31 families were documented and identified as liver and gallbladder herbal remedies in Northern Peru. Most species used were Asteraceae (9 species, 17.66%) followed by Euphorbiaceae (4 species, 7.85%) and Gentianaceae (3 species, 5.89%). All other families contributed only one or two species each to the pharmacopoeia. Asteraceae are clearly over-represented in comparison to the overall medicinal flora, while some other medicinally important families (e.g., Solanaceae, Lycopodiaceae, Cucurbitaceae, Rosaceae) are completely missing from the liver ailment category (Bussmann and Sharon 2006b).

The majority of herbal preparations employed for liver ailments were prepared from the whole plant (35.38%) while the leaves (24.61%), flowers (9.23%), and stems (12.32%) were used less frequently. Whole plants were more often used than characteristic for the overall medicinal preparations found in the region, while stems of plants were employed much less frequently (Bussmann and Sharon 2006b). This indicates that the local healers have a less well-developed knowledge about the constituents of individual plant parts in the case of liver and gallbladder treatments than for other applications. In almost 65% of the cases fresh plant material was used to prepare remedies, which differs little from the average herbal preparation mode in Northern Peru. Most of the remedies were applied orally (over 90%), while the remaining ones were applied topically. This differs a great deal from the regional average of application. Over 71% of all remedies were prepared as mixtures with multiple ingredients by boiling plant material either in water or in sugarcane alcohol. This indicates that the local healers have a very profound knowledge about the synergistic effects of plants in multi-ingredient preparations.

Almost no scientific evidence exists to date to prove the efficacy of the species employed as liver and gallbladder remedies in Northern Peru. Only 8% of the plants found or related species in the same genus have been studied.

Diarrhea, Stomach Problems, and Other Intestinal Ailments

Food-borne diseases are a serious public health problem worldwide. Some food-borne diseases are well recognized, but have recently become more common. Outbreaks of salmonella have been reported for decades, but within the past 25 years the disease has increased in incidence on many continents. While cholera has devastated much of Asia and Africa for years, its introduction for the first time in almost a century on the South American continent in 1991 makes it another example of an infectious disease that is both well-recognized and emerging. While cholera is often waterborne, many foods also transmit infection. Infection with *Escherichia coli* serotype O157:H7 (*E. coli*) was first described in 1982. Subsequently, it has rapidly emerged as a major cause of bloody diarrhea and acute renal failure. Outbreaks of infection, generally associated with beef, have been reported in Australia, Canada, Japan, the United States, in various European countries and in southern Africa (WHO 2002).

In total, 75 plant species belonging to 62 genera and 39 families were documented and identified as herbal remedies for intestinal ailments in Northern Peru. Most species used were Lamiaceae (13.33%) followed by Asteraceae and Rutaceae (both 5 species each, 6.67%). Most other families contributed only one species each to the pharmacopoeia. The most important anti-infectious families are clearly over-represented in comparison to the overall medicinal flora, while some other medicinally important families (e.g., Asteraceae) are much less important (Bussmann and Sharon 2006b).

The majority of anti-infectious herbal preparations were prepared from the leaves of plants (29.25%), the whole plant (22.64%), and stems (16.04%). This indicates that the local healers count on a very well developed knowledge about the properties of different plant parts. In almost 60% of the cases fresh plant material was used to prepare remedies, which differs little from the average herbal preparation mode in Northern Peru. Interestingly, only about 83% of the remedies were applied orally, while the remaining ones were applied topically. Over half of all remedies were prepared as mixtures of multiple ingredients by boiling plant material either in water or in sugarcane alcohol.

Large parts of the species used for intestinal disorders in Northern Peru are introductions from other parts of the world, especially Europe. Many of these are well known, and almost 50% of the plants found in this study have shown efficacy in scientific studies.

Reproductive Problems and Female Health

According to WHO (1999) estimates, reproductive problems - including 340 million new cases of curable Sexually Transmitted Diseases (STIs) such as syphilis, gonorrhoea, chlamydia, and trichomoniasis - occur annually throughout the world in adults aged 15-49 years of age. In developing countries, STIs and their complications rank in the top five disease categories for which adults seek healthcare. Infection with STIs can lead to acute symptoms, chronic infection and serious delayed consequences such as infertility, ectopic pregnancy, cervical cancer, and the untimely death of infants and adults (WHO 2007).

In total, 105 plant species belonging to 91 genera and 62 families were documented and identified as herbal remedies for reproductive problems in Northern Peru. Most species used were Asteraceae (9.52%) followed by Lamiaceae and Fabaceae (8.57% and 6.67%). Other families were less important, and 44 contributed only one species each to the pharmacopoeia. The most important families are represented in a manner similar to their overall importance in the local pharmacopoeia (Bussmann and Sharon 2006b).

The majority of herbal preparations for reproductive issues were prepared from the leaves of plants (22.72%), the whole plant (21.97%), and stems (21.21%), while other plant parts were used much less frequently. This indicates that the local healers count on a very well developed knowledge about the properties of different plant parts. In almost 62% of the cases fresh plant material was used to prepare remedies, which differs little from the average herbal preparation mode in Northern Peru. Over 70% of the remedies were applied orally, while the remaining ones were applied topically. Many remedies were prepared as mixtures of multiple ingredients by boiling plant material either in water or in sugarcane alcohol.

Little scientific evidence exists to prove the efficacy of the species employed as reproductive remedies in Northern Peru. Only 34% of the plants found or their congeners have been studied at all for their medicinal properties.

Heart and Circulatory System

Cardiovascular diseases are collectively the number one cause of death on the globe, accounting for over 30% of all deaths worldwide, 80% of which occur in lower income countries with often little Western healthcare available. Lower income groups generally have a higher prevalence of risk factors (WHO 2009). Traditional Medicine is used globally and has rapidly growing economic importance. In developing countries, traditional healers are frequently consulted to treat heart problems and disorders of the circulatory system. The healers encountered in this study use a wide variety of terms relating to heart problems which in part generalize the conditions (e.g., "heart disease"), including references to conditions as underlying causes of heart problems (e.g., "cholesterol") or simply using terms to indicate treatment options (e.g., "blood irrigation" as a term referring to "thinning" a patient's blood, "blood purification" or "refreshing the heart," a term indicating a process of cleansing the blood from suspected toxins and "blood circulation," indicating a treatment that improves circulation). The use of Western-style biomedical terms is not surprising, given that all informants are of Mestizo origin, and live in an urban environment.

Most treatments of the circulatory system involve the purification of the blood in order to improve the general condition of the patient. In addition, the fashionable concept of "weight management" and conditions related to obesity have entered the domain of Peruvian healers. All healers readily acknowledge the negative influence of high cholesterol levels and plant remedies were used specifically to lower cholesterol as well as weight loss therapies, while plants used for weight gain are insignificant.

In total, 60 plant species belonging to 52 genera and 33 families were documented and identified as heart herbal remedies in Northern Peru. Most species used were Asteraceae (7 species, 11.67%), followed by Lamiaceae (6 species, 10%) and Solanaceae (4 species, 6.67%). Fabaceae, Amaranthaceae and Cucurbitaceae each contributed 3 species (5%) to the heart pharmacopoeia. All other twenty-seven families contributed only one or two species each to the pharmacopoeia. Asteraceae are in general under-represented as heart remedies in comparison to the medicinal flora used in Northern Peru; Lamiaceae and Euphorbiaceae are clearly over-represented in comparison to the overall medicinal flora, while some other medicinally important families (e.g., Poaceae, Lycopodiaceae, Rosaceae) are completely missing from the heart category (Bussmann and Sharon 2006 a, b).

The majority of heart remedies were prepared from whole plants (37.18%), while the leaves (24.36%), stems (15.38%), and flowers (7.69%) were used less frequently. Whole plants were more often used than characteristic for the overall medicinal preparations found in the region (Bussmann and Sharon 2006b). In almost 70% of the cases fresh plant material was used to prepare remedies, which differs little from the main herbal preparation mode in Northern Peru. Over 90% of the remedies were applied orally, while the remaining ones were applied topically. This is very different from the regional average of application. Over 65% of all remedies were prepared as mixtures with multiple ingredients by boiling plant material either in water or in sugarcane alcohol. This indicates that the local healers have a very profound knowledge about the synergistic effects of plants in multi-ingredient preparations.

Little scientific evidence exists to date to prove the efficacy of the species employed as heart remedies in Northern Peru. Only 33% of the plants found or related species in the same genus have been studied at all.

Inflammation and Bacterial Infections

Bacterial infections and inflammation are among the ailments responsible for a large number of deaths worldwide and are often treated by traditional healers (Bussmann and Sharon 2006a, b; WHO 2009).

In total, 96 plant species belonging to 84 genera and 46 families were documented and identified as anti-infectious herbal remedies in Northern Peru. Twenty percent of the species were introductions, while 80% belong to the native flora of Peru. Most species used belong to Asteraceae (18.95%), followed by Fabaceae and Euphorbiaceae (7.37% and 5.26%, respectively). Most other families contributed only one species each to the pharmacopoeia. The most important anti-infectious families were over-represented in comparison to the overall medicinal flora, while some other medicinally important families (e.g., Lycopodiaceae, Cucurbitaceae) are completely missing from the anti-infectious category.

The majority of herbal preparations were prepared from the leaves of plants (31.34%), while the whole plant (18.66%), flowers (12.69%), and stems (17.16%) were used less frequently. In almost 67% of the cases fresh plant material was used to prepare remedies. Only about 55% of the remedies were applied orally, while the remaining ones were applied topically. Over half of all remedies were prepared as mixtures of multiple ingredients by boiling plant material either in water or in sugarcane alcohol. Infections, in particular by strains of *Staphylococcus aureus* are very common and increasingly difficult to treat due to widespread formation of drug resistance. Fungal infections, due to the structure of the organisms involved, have always been difficult to treat. Given the high importance of infections, it is not surprising that anti-infectious agents are high on the list for drug development, and a large number of species used traditionally have undergone screening. Almost 43% of the plants used in Northern Peru to treat infections or their congeners have been studied for their medicinal properties.

Malaria and Fever

Malaria is still a major global public health problem in most tropical countries. It is thought that malaria is by far the most serious tropical disease causing one to two million deaths per year, and it plays a major role in the high mortality seen in infants and children (El Kamali et al. 1997; Milliken 1997). It is also responsible for miscarriages, premature deliveries, growth retardation, low birth weight, and anemia (Connally 1996; Gbile 1984; Minakawa et al. 2002; Hay et al. 2003).

The World Health Organization (WHO) has estimated that about 2 billion people in over 100 countries are exposed to malaria, with 247 million cases in 2006 alone, and half of the world's population is potentially exposed to the disease (WHO 2009). The worsening global economic situation makes it difficult to expand modern health services, hence an effective low-cost delivery medical system is urgently needed (El Kamali et al. 1997).

This is even more pressing because the use and misuse of over-the-counter anti-malarial remedies like chloroquine to prevent and treat *falciparium* malaria has led to widespread appearance of resistant parasites (Milliken 1997). This is complicated by the fact that global warming may lead to expansion of areas in which the ambient temperature and climatic conditions are suitable for *Plasmodium* transmission. Climatic variability has been associated with some of the recent epidemics (Minakawa et al. 2002).

In total, 17 plant species belonging to 17 genera and 13 families were documented and identified as anti-malarial herbal remedies in Northern Peru. Most species used were Asteraceae (3 species, 17.66%), followed by Fabaceae, and Solanaceae (both 2 species each, 11.77%). All other families contributed only one species each to the pharmacopoeia. The most important anti-malarial families are clearly over-represented in comparison to the overall medicinal flora, while some other medicinally important families (e.g., Lamiaceae, Euphorbiaceae, Poaceae, Apiaceae) are completely missing from the anti-malarial category (Bussmann and Sharon 2006b). In the context of the surveys, healers and vendors often referred to "fever" when talking about malaria. Fever however included a variety of conditions, from fever accompanying flu to fever as a result of malaria. Malaria was recognized as a parasitic infection, and treated accordingly, while other plant species were used to treat fever as a symptom, mainly focusing on lowering body temperature.

The majority of anti-malarial herbal preparations were prepared from the leaves of plants (38.46%), while the whole plant (26.92%), flowers (15.38%), and stems (11.54%) were used less frequently. Leaves and stems were used more often for malaria treatments than would have been expected in comparison to the overall medicinal preparations found in the region, while the seeds of plants were employed much less frequently and other plant parts not at all (Bussmann and Sharon 2006b). This indicates that the local healers count on a very well developed knowledge about the properties of different plant parts. In almost 70% of the cases fresh plant material was used to prepare remedies, which differs little from the average herbal preparation mode in Northern Peru. Interestingly, about 55% of the remedies were applied orally, while the remaining ones were applied topically. This differs little from the regional average of application. Over half of all remedies were prepared as mixtures of multiple ingredients by boiling plant material either in water or in sugarcane alcohol.

The very limited number of plants employed on the Peruvian coast to treat malaria and fevers might at first glance seem surprising if compared to studies from other regions of the country (Kvist et al. 2006; Roumy et al. 2007). However, malaria has always been of relatively minor importance in the coastal desert areas. Thus it is not surprising that few remedies are employed. There are indications that health practices are in the process of changing, and traditional healers are starting to treat patients with prepared Western remedies (e.g. aspirin, primaquin, malaraquin or lariam) although plant preparations are still important (Bussmann and Sharon 2006b, 2007b, 2009a).

Little scientific evidence exists to prove the efficacy of the species employed as malaria remedies in Northern Peru. Only 41% of the plants found or their congeners have been studied for their medicinal properties.

Cancer and Diabetes

Fourty-seven plant species belonging to 42 genera and 30 families were used by *curan-*

deros in Northern Peru to treat cancerous conditions and diabetes symptoms. Most species used were Asteraceae (9 species, 19.15%), followed by Gentianaceae (3 species, 6.37%) and seven familes with two species each (4.25%). All other families contributed only one species each to the pharmacopoeia. Asteraceae as the most important anti-cancer and anti-diabetic family is clearly over-represented in comparison to the overall medicinal flora, while most other medicinally important families are either under-represented or completely missing from the category (Bussmann and Sharon 2006b).

The majority of anti-cancer and anti-diabetic herbal preparations were prepared from the leaves of plants (30.77%), while the whole plant (20%), stems (20%), and flowers (6.15%) were used less frequently. Leaves and stems were more often used than characteristic for the overall medicinal preparations found in the region, while whole plants were employed less frequently (Bussmann et al. 2006b). This indicates that the local healers count on a very well developed knowledge about the properties of different plant parts. In almost 60% of the cases fresh plant material was used to prepare remedies, which differs little from the average herbal preparation mode in Northern Peru. Over 90% of the remedies were applied orally, while the remaining ones were applied topically. This is significantly different from the regional average of application. More than 50% of the remedies included multiple plants.

Little scientific evidence exists to date to prove the efficacy of the species employed as anti-cancer and anti-diabetic remedies in Northern Peru. Only 38.71% of the plants found as diabetes treatments and 17.65% employed as anti-cancer remedies or related species in the same genera have been studied.

Parts of Medicinal Plants Used and Mode of Application

Northern Peruvian *curanderos* prefer to use either the leaves (in 25% of all uses) or the whole plant (24%) for the preparation of their remedies. In 19% of the cases the stems of the plants were used, most commonly together with the leaves. Flowers (10%), seeds (7%), fruits and roots (4% each), bark (3%), fruit peel (2%), and latex and wood (1% each) were used for a small number of preparations.

Almost two-thirds (64%) of the remedies employed in Northern Peru are prepared using fresh plant material. Many of the introduced species are cultivated in fields and gardens, but the majority of the indigenous species are collected wild. This indicates that a widespread system of plant collectors is needed to supply the fresh plant material needed in Traditional Medicine. Most healers agreed, however, that in most cases dried material could be used if fresh plants were not available. In 36% of all cases the remedies were prepared using specifically dried plant material. Fresh material was not used in these situations (Bussmann and Sharon 2006b).

Healers in Northern Peru often employ very sophisticated mixtures of a variety of plants in their treatments. The use of single species for treatments was rare. Most commonly, plant material was boiled in water, or in some cases in sugarcane alcohol (*aguardiente*) to extract the active compounds. In some cases, plant material was macerated in cane alcohol or wine for longer periods of time before use.

The *curanderos* all had strikingly exact recipes for treatment, with very specific quantities of plant material used to prepare remedies. These quantities did not differ greatly from one healer to another. Also, the amount of a specific remedy that was given to a patient was very similar among the different *curanderos.*

The most frequent way to administer remedies was to prepare a decoction and ingest it orally (52% of all uses) followed by application as a poultice (38%, plant crushed and/or boiled and applied). Seven percent of all plant uses entailed the preparation of a *seguro*, a bottle or small flask filled with plant material along with various perfumes. This amulet has to be carried by the patient at all times, or it is placed in the house and used for periodic blessings. *Seguros* contained anything from a handful to more than three dozen different ingredients. In two percent of the plant uses the material was employed to fabricate charms, and in one percent of all applications the plant material was burned as incense, with the smoke inhaled for treatment.

Many traditional healers rely on herbal preparations - often consisting of complex ingredients and with very specific preparations - to treat their patients' illnesses, rather than just employing single plant extracts. However, studies documenting these preparations and analyzing the composition of the mixtures are almost non-existent. Most ethnobotanical studies to date document the "use" of single species, without asking the important question if the plants in question are really employed alone, or if they are in fact part of a more complex preparation. Cano et al. (2004) were amongst the first authors to respond to this challenge, reporting on plant mixtures employed in Cuba and the Middle East, while Vandebroek et al. (2010) demonstrated the great complexity of plant preparations in the Dominican Republic. However, no information was available for the very species-rich Andean pharmacopoeia.

The present publication attempts to give a detailed overview on the herbal mixtures employed by traditional practitioners in Northern Peru and the specific applications for which they are used in order to provide a baseline for more in-depth studies on efficacy and safety of these preparations, as well as the possible applications in the public health system.

The investigation of plant mixtures used in traditional medicine in Northern Peru yielded a total of 974 herbal preparations used to treat 164 different afflictions (Bussmann, Glenn, Meyer, Kuhlman and Townesmith 2010). The classification of diseases followed *curandero* terminology. To allow a better overview the different disease concepts were grouped in more inclusive disease categories according to their similarities. Psychosomatic disorders were the most outstanding afflictions treated with traditional herbal mixtures, with almost 30% of all recipes applied, followed by respiratory illnesses, female issues, kidney problems and heart problems. *Susto* (fright), problems of the nervous system, general systemic inflammation and bronchitis together accounted for almost 25% of all remedies used. In many cases, healers used only one or two common mixtures to treat an illness. This degree of consensus between different healers shows great sophistication in the diagnosis and treatment of specific disorders. On the contrary, when it came to the treatment of nonspecific disease categories like "inflammation" or "bronchitis," every healer seemed to use her/his own specific mixture to treat the problem. This was particularly obvious in the treatment of neurological and psychosomatic problems, for which the majority of plants and mixtures was employed. Up to 49 different preparations were used to treat the same disease. This seems to indicate a high degree of ongoing experimentation in order to find working cures for nonspecific symptoms and that there is very little consensus amongst the individual healers as to which cure to employ. This low consensus, especially where spiritual and nervous system/psychosomatic disorders are involved, might also indicate that the individual healers are reluctant to exchange knowledge about their specific and protected treatment methodologies for these categories, while the knowledge about "simple" treatments is much more widespread.

Altogether 330 plant species, representing almost 65% of the medicinal flora used in the region (Bussmann et al. 2010a) were applied in mixtures. Of these, 64 species (19.39%) were introductions, falling within the range of introduced species as a percentage of all the medicinally applied flora. Among the plants employed, Asteraceae stood out as expected, and the number of species used in this family was comparable to the percentage of Asteraceae in the medicinal flora of the region (Bussmann and Sharon 2006). The overwhelming number of plant mixtures contained two to seven different plant species, although in the most extreme case 27 distinct species were included. A large number of species appeared in various mixtures. The plant species for each mixture are listed in the order given by the *curanderos* so as to express the importance of the individual species, rather than providing an alphabetical listing. (For a detailed overview on quantities and parts of each plant used see Bussmann and Sharon 2006b).

Cluster analysis confirmed that mixtures used for applications like inflammations, infections and blood purification as well as cough, cold, bronchitis or other respiratory disorders, or urinary infection and kidney problems had similar floristic compositions. However, a few interesting clusters stood out, e.g., mixtures used for nervous system disorders, anxiety and heart problems often had a similar composition as did mixtures for prostate and bladder problems; kidney problems, gallbladder disorders, diabetes, and cholesterol were treated with the same preparations as were rheumatic illnesses and asthma. Our research suggests that this indicates that the local healers have a very detailed understanding of disease concepts and are choosing their remedies very carefully, based on the diagnosed cause, i.e., heart problems get treated differently if they are caused by stress versus a physical agent while kidney infections are treated differently from kidney problems linked to diabetes and/or obesity.

The floristic composition as well as the complex phytochemistry of traditional herbal mixtures remains woefully understudied. This is all the more surprising since traditional one-plant/single-compound based drug discovery efforts have yielded very few results in recent decades, which might in fact provide an explanation for the fact that so many plant species that have been documented for a certain use are found to be "inefficient" or "toxic" when subjected to clinical trials.

Our research indicates that a large number of plants used in traditional healing in Northern Peru are employed in sophisticated mixtures, rather than as individual plants. Peruvian *curanderos* appear to employ very specific guidelines in the preparation of these "cocktails" and seem to have a clear understanding of disease concepts when they diagnose a patient. This in turn leads them to apply specific mixtures for specific conditions. There seems to be a widespread exchange of knowledge about mixtures for treatment of bodily diseases, while mixtures for spiritual, nervous system, and psychosomatic disorders appear to be more closely guarded by the individual healers.

Traditional herbal mixtures, with their wealth of compound fragments and new compounds originating in the preparation process, could well yield clues to the treatment of a wide variety of diseases. The present book provides detailed baseline information on the composition and use of traditional mixtures in Northern Peru. Further studies to compare the compound composition of these preparations versus single-plant extracts, as well as investigations comparing efficacy and toxicity of herbal preparations in comparison with their single-plant ingredients are in progress.

Does Traditional Medicine Work? A Look at Antibacterials Used in Northern Peru

Plants with potential medicinal activity have recently come to the attention of Western scientists, and studies have reported that some are bioactive (Perumal Samy et al. 2000). Potentially active compounds have been isolated from a few of the plants tested (D'Agostino et al. 1995 a, b; Okuyama et al. 1994, Rodriguez et al. 1994).

In order to evaluate the antibacterial activity of species used in Traditional Medicine in Northern Peru, 525 plant samples of at least 405 species were tested using simple agar-bioassays for antibacterial activity against *Staphylococcus aureus, Escherichia coli, Salmonella enterica Typhi* and *Pseudomonas aeruginosa.* A much larger number of ethanol plant extracts showed antibacterial activity compared to water extracts. One hundred ninety-three ethanol extracts and 31 water extracts were active against *S. aureus.* In 21 cases only the water extract showed activity (for all bacterial species) compared to ethanol alone. None of the aqueous extracts were active against the other three bacteria, with the activity of the ethanol extracts also much reduced since only 36 showed any activity against *E. coli* and 3 each against *S. enterica Typhi* and *P. aeruginosa.* Eighteen ethanol extracts were effective against both *E. coli* and *S. aureus,* while in two cases the ethanol extract showed activity against *E. coli* and the water extract against *S. aureus.* The ethanol extract of *Dioscorea trifida* was effective against *E. coli, S. aureus* and *P. aeruginosa. Caesalpinia spinosa* was the only species that showed high activity against all bacteria, including *Salmonella enterica Ttyphi* and *Pseudomonas aeruginosa* when extracted in ethanol (Bussmann et al. 2010).

Two hundred twenty-five extracts came from plant species that are traditionally employed against bacterial infections. One hundred sixty-six (73.8%) of these were active against at least one bacterium. Of the three hundred extracts from plants without traditional antibacterial use, only 96 (32%) showed any activity. This clearly demonstrates that plants traditionally used as antibacterial agents had a much higher likelihood of being antibacterially active than plants without traditional anti-bacterial uses. However, the efficacy of plants used traditionally for antibacterial related applications did vary, which underscores the need for studies aiming at clearly understanding traditional disease concepts.

Plants used for respiratory disorders, inflammation/infection, wounds, diarrhea and the prevention of post-partum infections were efficacious in 70-88% of the tests. Plants used for "kidney inflammation" had a much lower efficacy against bacteria, falling within the range of species that are traditionally used to treat other bodily disorders. Only species used for spiritual/ritual treatments scored worse. Of these only 22% showed some antibacterial activity. However, amongst the "spiritual" plants 38% of the species used for cleansing baths did in fact show activity, while only 15% of the plants often used in protective amulets (mostly species within the families of Lycopodiaceae and Valerianaceae) showed limited antibacterial activity.

Several species showed higher efficacy than the control antibiotics employed. For example, *Ambrosia peruviana, Iresine herbstii, Niphogeton dissecta, Opuntia ficus-indica,* and *Smilax kunthii* were particularly effective against *Escherichia coli. Berberis buceronis, Caesalpinia paipai, Caesalpinia spinosa, Cestrum strigilatum, Cydista aequinoctialis, Dioscorea trifida, Escallonia pendula, Escobedia grandiflora, Eucalyptus citriodora, Eucalyptus globulus, Eugenia obtusifolia, Eustephia coccinea, Gallesia integrifolia, Geranium sessiliflorum, Hedyosmum racemosum, Iresine herbstii, Lycopersicon hirsutum, Mauria heterophylla, Phyllanthus niuriri, Porophyllum ruderale, Salvia cuspidata, Senecio chionogeton, Smilax kunthii, Tagetes erecta,* and *Taraxacum officinale* showed high activity against *Staphylococcus aureus.* The same holds true for *Ephedra americana, Gentianella bicolor,* and *Mandevilla* cf. *trianae.* However, extracts of these three species were highly inconsistent in their efficacy.

The comparison of closely related species traditionally employed for different purposes (e.g, different *Alternanthera* spp., *Passiflora* spp., *Senecio* spp, and *Salvia* spp., for spiritual purposes and against bacterial infections) showed that the "spiritual" species normally were not effective against bacteria, while the species used as antibacterials were effective. The example of *Plantago sericea* var. *sericea* (used in *seguros* with no efficacy) and *Plantago sericea* var. *lanuginosa* (used for vaginal infections with high efficacy against *S. aureus*) are particularly compelling cases that indicate the sophistication of traditional knowledge. However, we did find examples like *Chuquiragua* spp. where closely related species were used as antibacterials, but only one of them did in fact show efficacy, clearly indicating that in this case traditional knowledge did not produce reliable results.

On the other hand, extracts of the same species traditionally used to treat infections often produced vastly diverging results when collected from different localities. Good examples are *Iresine herbstii, Schinus molle, Eustephia coccinea, Oreopanax eriocephalus, Myroxylum balsamum, Spartium junceum,* and *Gentianella dianthoides.* Most of these species did not produce particularly high inhibition rates and were not the first choice of healers when trying to find remedies for bacterial infections. Many traditional remedies for concepts like "kidney inflammation" did not produce any antibacterial results, which indicates that research into efficacy does need to take traditional disease concepts into account.

Many remedies used for spiritual healing and other non-infection purposes did show antibacterial efficacy in vitro, but were not described as such by the local healers. This might be explained by the fact that they either are very inconsistent in their activity (e.g., *Mandevilla trianae, Loricaria* spp., *Lonicera japonica, Hypericum laricifolium, Hyptis sidifolia, Mentha piperita, Brachyotum naudinii, Cydonia oblonga*) or are so closely related that identification, especially when dried, can be a problem, e.g., in the case of *Baccaris* spp., *Gentianella* spp., and *Valeriana* spp. Or they may be prone to toxic side effects like *Ephedra americana* and *Brugmansia* spp.

Almost all remedies are traditionally prepared as water extracts, although ethanol (in the form of sugarcane alcohol) is readily available. This might at a first glance seem astonishing, given the low efficacy of water extraction found in this study. However, initial results from brine-shrimp toxicity assays indicate that the ethanol extracts are far more toxic than water extracts of many species. Thus ethanol extraction might in many cases not be suitable for therapeutic application. This again indicates the considerable sophistication and care with which traditional healers in Northern Peru choose their remedies for specific purposes.

If the botanical documentation of Peruvian medicinal plants has been neglected, investigations of the phytochemical composition of useful plants is lagging even further behind. Most studies on the phytochemistry of Peruvian plants concentrate on a few "fashionable" species that have been marketed heavily on a global scale, especially Maca (*Lepidium meyenii*), Sangre del Drago or del Grado (*Croton lechleri*) and Uña de Gato (*Uncaria tomentosa* and *Uncaria guianensis*). The number of other Peruvian plants for which at least some phytochemical studies exist is still miniscule, and most efforts are fuelled by the fads and fashions of the international herbal supplement market. Studies involving multiple species were only initiated as late as the 1990s.

Minimum inhibitory concentrations for Peruvian plant extracts ranged from 0.008 to 256mg/ml. The high values in many species indicate a very limited antibacterial efficacy. The ethanol extracts exhibited stronger activity and a much broader spectrum of action than the water extracts. The most interesting activity with *E. coli* was obtained from ethanol extracts of *Baccaris* sp., *Ochroma pyramidale, Croton lechleri, Banisteriopsis caapii, Miconia salicifolia,* and *Eugenia obtusifolia.* Only the latter species also showed strong activity in the aqueous extract. A much wider range of species, including most species active against *E. coli* showed inhibition of *S. aureus. Poropohyllum ruderale, Senecio* sp., *Corynaeae crassa, Dioscorea trifida, Senna monilifera, Spartium junceum, Pelargonium odoratissimum, Satureja pulchella, Cuphea* sp., *Malva parviflora, Brosmium rufescens, Syzygium aromaticum, Sanguisorba minor, Citrus limetta, Verbesine* sp., and two unidentified species all showed MIC values between 1-4mg/ml. Most of them however did not display any efficacy in aqueous extract. *Hypericum laricifolium, Hura crepitans, Caesalpinia paipai, Cassia fistula, Hyptis sidifolia, Salvia* sp., *Banisteriopsis caapi, Miconia salicifolia,* and *Polygonum hydropiperoides* showed the lowest MIC values and would be interesting candidates for future research (Bussmann et al. 2010).

Most species effective against *S. aureus* are traditionally used to treat wound infections, throat infections, serious inflammations or post partum infections. Interestingly, many species used in cleansing baths also showed high activity against this bacterium. Many of these species are either employed topically or in synergistic mixtures so that possible toxicity seems not to be an issue. The species effective against *E. coli* were mostly employed for conditions that traditional healers identified as "inflammation."

Most of the plants used by the healers have antibacterial activity, but only seven of the 141 plants (5.6%) examined in this study show any MIC values of 200 or less mg/ml of extract. Of these plants, five are used to treat diseases believed to be bacterial in origin by Traditional Medicine, one is a disease not believed to be caused by bacteria, and one is used for undefined treatment purposes.

Nine out of 141 plants (6.3%) tested were not used for diseases believed to be bacterial in origin by Traditional Medicine. Five showed high antibacterial activity with MIC values below 16 mg/ml. Four of these were among the most potent plants tested with MIC values of two or less mg/ml including a hallucinogen and extracts used to treat diabetes and epilepsy. Diseases such as diabetes often compromise the health of the individual and antibacterial treatments may be warranted for secondary complications of the disease.

Toxicity in Traditional Medicine

Basic medicinal activities have been investigated for a wide variety of plants. However, while toxicity assays are available for a large number of countries, no data exist on the potential toxicity of Peruvian medicinal species.

Brine shrimp *(Artemia)* are frequently used as an agent in laboratory assays to determine toxicity values by estimating LC_{50} values (median lethal concentration). The brine shrimp lethality activity of 501 aqueous and ethanol extracts of 341 plant species belonging to 218 genera of 91 families used in Peruvian Traditional Medicine was tested (Bussmann, Malca, Glenn et al. 2011). The aqueous extracts of 55 species showed high toxicity values (LC_{50} below 249 mg/ml). Eighteen species showed median toxicity (LC_{50} 250-499 mg/ml) and 18 showed low toxicity (LC_{50} 500-1000 mg/ml). The ethanol extracts proved to be much more toxic: 220 species showed high toxicity values (LC_{50} below 249 mg/ml, with 37 species having toxicity levels of >1mg/ml), 43 species showed median toxicity (LC_{50} 250-499 mg/ml), and 23 species showed low toxicity (LC_{50} 500-1000 mg/ml). Over 24% of the aqueous extracts and 76% of the ethanol extracts showed elevated toxicity levels to brine shrimp. Traditional preparation methods are taking this into account. Most remedies are prepared as simple water extracts, thus avoiding potential toxic effects. Excellent examples occur where the water extracts are non-toxic, while the ethanol extracts showing high toxicity are *Ocimum basilicum* L., *Salvia* sp. or *Laccopetalum giganteum* (Wedd.) Ulbrich. In contrast, *Cinchona officinalis* L. ethanol extracts were non-toxic and are used traditionally, while the highly toxic water extract has no traditional use.

Species which showed higher levels of toxicity were *Bejaria aestuans* L., *Erodium cicutarium* (L.) L'Her., *Brachyotum naudinii* Triana, *Miconia salicifolia* (Bonp. ex Naud.) Naud., *Cuscuta foetida* Kunth, *Caesalpinia spinosa* (Molina) Kuntze, and *Phyllactis rigida* (Humb. and Bonpl.) Pers. *Achillea millefolium* L., *Artemisia absinhitum* L, and *Eucalyptus globulus* Labill, frequently used as medicinal teas, also fall into this group as do *Lupinus mutabilis* Sweet and *Ilicium verum* Hook. f. Solanaceae (e.g., *Nicotiana tabacum* L. and *Solanum americanum* Mill.) proved highly toxic, while other species known to be highly toxic when ingested (e.g., *Datura* sp. and *Brugmansia* spp.) did not show toxicity in brine shrimp.

Multiple extracts from different collections of the same species in most cases showed very similar toxicity values. However, in some cases the toxicity of extracts from different collections of the same species varied from non-toxic to highly toxic. Examples for such variation in toxicity were found for *Chersodoma deltoidea* M.O. Dillon and Sagast., *Satureja sericea* (C. Presl. and Benth.) Briq., *Eugenia obtusifolia* Cambess., *Epidendrum* sp., *Capparis crotonoides* Kunth, *Sambucus peruviana* Kunth, and *Malva* sp. For these frequently used species, harvest time, collection locality or use of specific plant parts might be important for a reduction of toxicity.

Markets and Sustainability

The Pharmacopoeae of Southern Ecuador and Northern Peru - Colonial Regimes and Their Influence on Plant Use

The differences in medicinal plant use between Southern Ecuador and Northern Peru are striking. Both regions share the same cultural background, and have a very similar flora with a comparable number of plant species that to a large extent overlap. Many plants mentioned in Martinez de Compañón (1789) are still found in local markets today (Fig. 6). However, the medicinal flora of Southern Ecuador includes only 40% of the species used in Northern Peru. The differences in traditional medicinal use can be explained by comparing the development of the pharmacopoeiae of both areas from the start of the colonial period until today. Colonial chroniclers often included detailed descriptions of useful plants in their reports. Monardes (1574), Acosta (1590), and Cobo (1653/1956) provided the most comprehensive early accounts of the economically interesting flora of Northern Peru and Southern Ecuador. Later treatments were included in Alcedo (1776). Martínez de Compañón (1787), Archbishop of Trujillo, had a complete inventory of his dioceses prepared. Finally, Ruiz and Pavón provided the first real botanical inventory of the region (1777-1788). The account of Martínez de Compañón provides the best baseline for a comparison of the colonial and modern medicinal flora of the region. The work includes detailed paintings for every species, which allows a close comparison with the modern medicinal flora, indicating that the vernacular names of useful plants have not changed significantly since colonial times. It contains 526 useful plant species. A preliminary review of this work seems to indicate that the number of plants used has not changed significantly since the late 1700's, with over 500 plant species still found in modern Peruvian markets. A closer comparison shows, however, that only 41% of the species mentioned by Martinez de Compañón (1789) are still sold today in Peru. An additional 32% are still used in the Amazon basin, but no longer reach the coastal markets. Twenty-seven percent have completely disappeared from modern-day use. This means that 59% of the species sold in Peruvian markets and 41% of the species used in Ecuador were added to the pharmacopoeia within the last 200 years (Fig. 7).

A cluster analysis of the colonial and modern plant inventories showed a striking explanation for the use differences between Ecuador and Peru and helps to explain why the plant inventories changed so significantly in the 18th century. The current pharmacopoeia of useful flora in Ecuador was most similar to the early colonial flora mentioned in Monardes (1574), Acosta (1590), Cobo (1653/1956) and Alcedo (1776). This indicated that the Ecuadorian medicinal flora did not develop much between early and late colonial times. In contrast, the modern Peruvian healing flora was much more similar to later collections (Fig. 8). An explanation for this lies in the different treatment of traditional practices in Ecuador and Peru. In Ecuador, traditional medicinal practitioners were immediately persecuted once the colonial administration took hold, while the Peruvian administration was much more tolerant. This also reflects the establishment of a National Institute for Traditional Medicine in Peru in the 1980s, while Traditional Medicine was illegal in Ecuador until a constitutional change in 1998. This meant that Ecuadorian healers had no opportunity to experiment with new species to cure diseases introduced by Europeans, while Peruvian healers were able to explore the rich flora of the region in order to find new remedies. This experimentation also extended to "magical"

Figure 6: Flor de San Juan from drawing in Martinez de Compañón (1789) and recent market purchase.

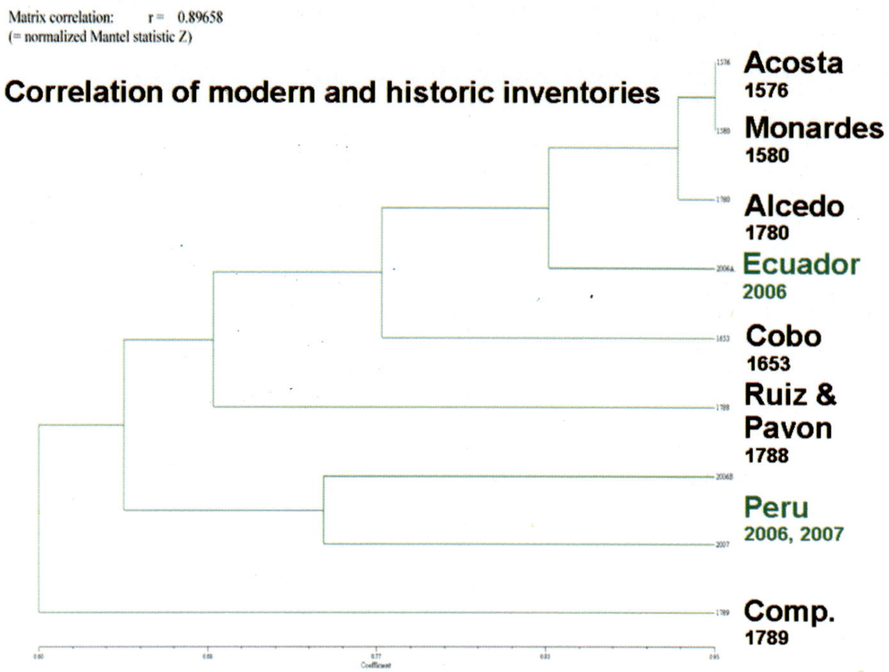

Figure 7: Cladogram of linkage of historic sources and current pharmacopoeiae

disease concepts like *Mal Aire, Mal Ojo, Susto,* and *Envidia* which were introduced from Spain during the colonial regime. Peruvian healers developed a vast array of medicinals to treat these conditions, which to a large extent explains the shift in the medicinal flora between the late 1700s and modern times. Experimentation in Ecuador remained restricted to the treatment of common diseases, while spiritual treatments were outlawed until the constitutional revision of 1998 recognized the right of the population to use Traditional Medicine (Fig. 9) (Bussmann and Sharon 2009a).

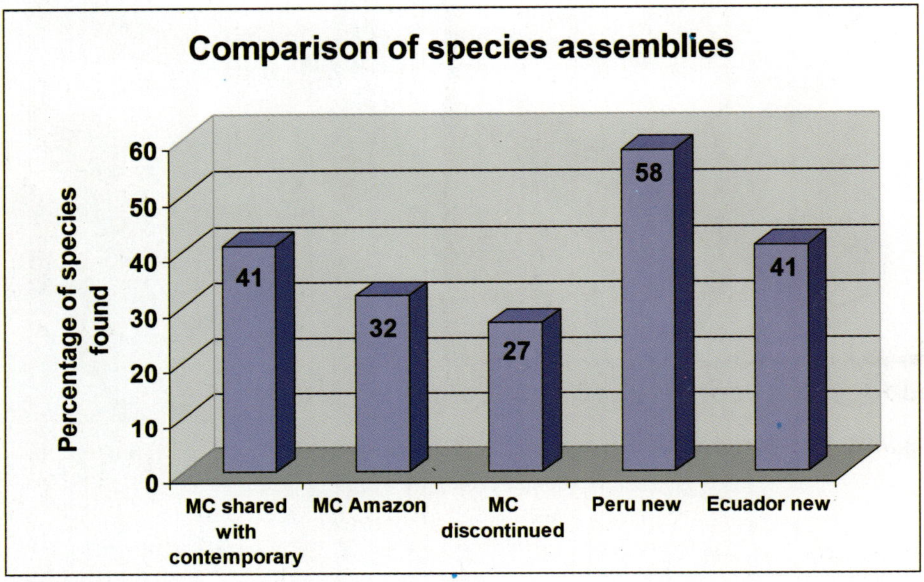

Figure 8: Differences in plant usage between Peru and Ecuador.

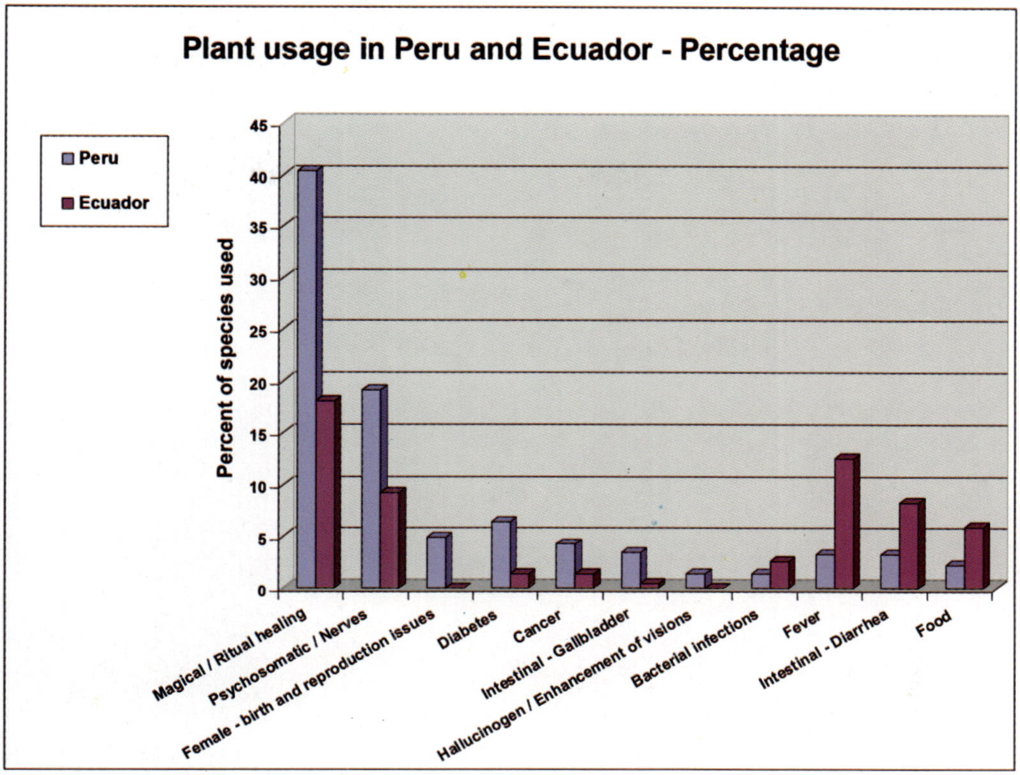

Figure 9: Plant use differences in Ecuador and Peru

The Case of *Ulluchu*

Ulluchu is the common name assigned to a plant frequently depicted in the art of the Moche culture, which thrived on the north coast of Peru from A.D. 100 to 800. It is a grooved, comma-shaped fruit with an enlarged calyx found mainly in fine-line scenes painted on Moche ceramics (Fig. 10). The term first appeared without linguistic explanation in the work of pioneer Moche scholar Rafael Larco Hoyle (1939: Fig. 58; 1940: 98, Figs. 166 & 167). In his 1939 publication, he reported that the peoples of the sierra and the coastal regions (Virú and Moche valleys) believed that the fruit had to be picked silently to prevent it from turning bitter. He wondered if the plant symbolized the silence and discretion of richly-attired Moche messengers, some of whom wear belts adorned with *Ulluchus*. In his 1938 publication, he labeled a Moche fine-line drawing of *Ulluchu* as *Phaseolus* sp. (a bean). Larco clearly recognized that *Ulluchu* had nothing whatsoever in common with "*ulluco*" (*Ullucus tuberosus*), an Andean tuber still widely cultivated and consumed in Peru today.

The symbolic importance of *Ulluchu* in Moche iconography was firmly established by Moche scholar Donna McClelland (1977). Based upon a meticulous review of the UCLA Moche Archive, she showed that its distribution was non-random and that its varied usage displayed definite patterns, with the greatest variability among background elements and the most frequent representations found on the belts of warriors and runners. She demonstrated that "the leaves of the *Phaseolus* do not resemble the *ulluchu* leaf depictions" (McClelland

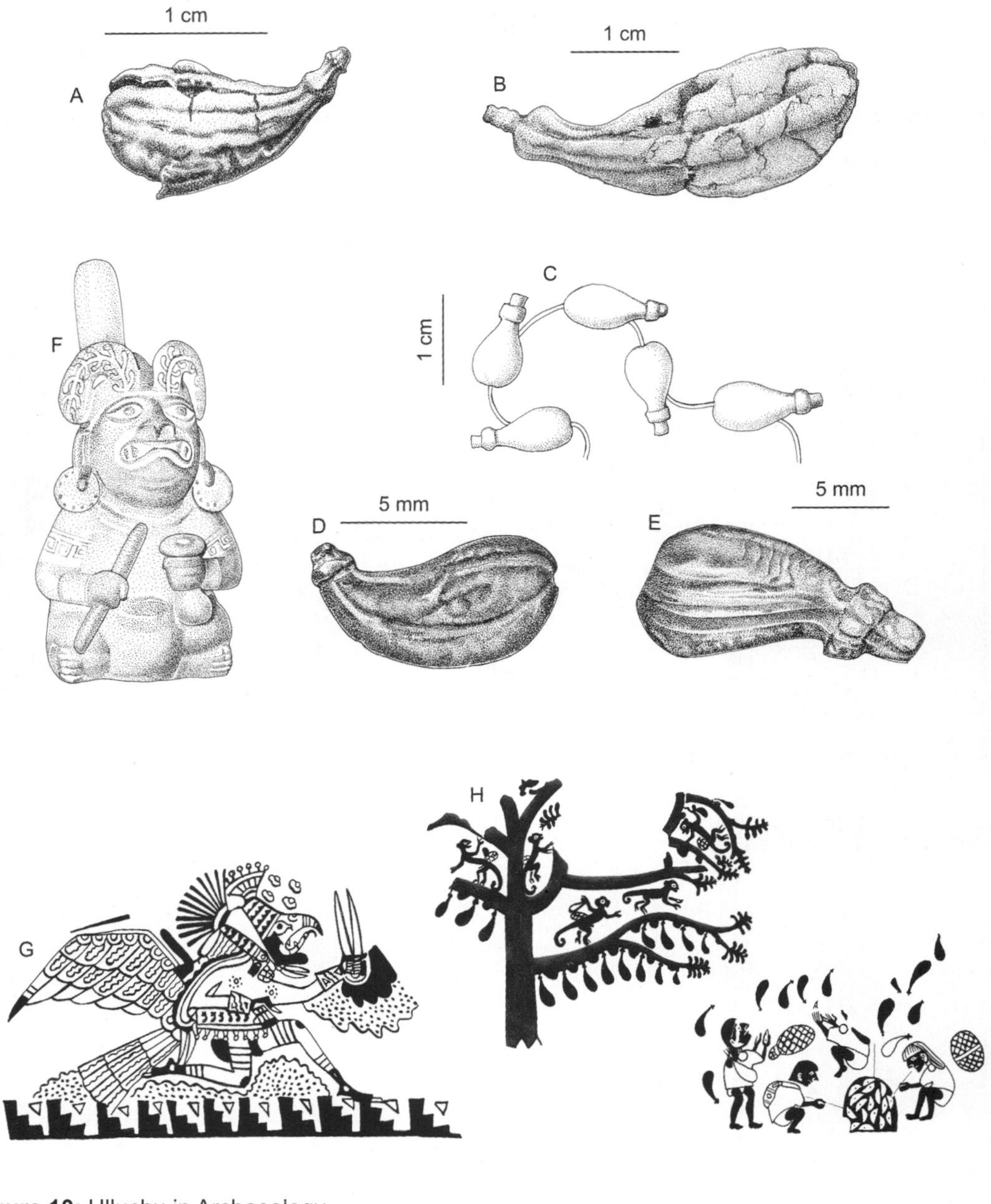

Figure 10: Ulluchu in Archaeology

A. Ulluchu fruit from cache at Sipán. After photograph by Christopher B. Donnan. B. Ulluchu fruit from Dos Cabezas burial. After photograph by Donald McClelland. C. Bone beads in form of Ulluchus fron Huaca de la Luna. After photograph by Donald McClelland. D. Golden Ullucho bead. After photograph by Donald McClelland. E. Spondylus shell bead in form of Ulluchu. After photograph by Donald McClelland. F. Supernatural figure seated holding a gourd, possibly containing ground Ulluchu seeds. Ulluchus painted on headdress. Private collection. After photograph by Christopher B. Donnan, from McClelland (2008). G. Anthropomorphized hawk runners with Ulluchu on belts and Ulluchu seeds floating above. Runner carrying snuff tube. The Art Institute of Chicago. After drawing by Donna McClelland (2008). H. Ulluchu harvest. Note tree with opposite leaves and extracted seeds on bottom right. After McClelland (2008), Private collection.

1977: 43). *Pepino (Solanum muricatum)* and *aji (Capsicum annum),* which are clearly depicted in Moche art and do not resemble *Ulluchu* were also eliminated "since the *ulluchi* [*sic*] fruit is suspended from the plant by its smaller pointed end, whereas these two are suspended by the large end" (McClelland 1977: 437). She also indicated that the plant had not been botanically identified, pointing out that, if it turned out to be a mythical plant, no identification would be possible.

A decade after McClelland's seminal article, S. Henry Wassen (1987) of the Gothenburg Ethnographical Museum, eliminated *Persea americana* Miller var. *americana* (a wild relative of *avocado*) as a candidate, concluding that *Ulluchu* was *Carica candicans* A. Gray (a species of wild *papaya*). He also co-authored an article describing the enzyme papain, which can be extracted from unripe papaya for use as a blood anti-coagulant (Hulten et al. 1987). In the latter article, the authors proposed that papain was used in the Moche sacrifice ceremony to prevent the coagulation of blood drawn from sacrificed warriors for later consumption by priests.

In a paper presented at the Sibley Conference at the University of Texas at Austin in November 2003, McClelland (2008) - in addition to updating her 1977 paper in the light of a vastly expanded Moche Archive and archaeological discoveries of real *Ulluchu* - refuted the papaya hypothesis. She also discussed the presence in the art of yellow oleander seeds (*Thevetia peruviana*) as legging rattles as well as *espingo* seeds (*Nectandra* sp.) which Wassen (1976) had earlier suggested might have been added to corn beer for medicinal and psychotropic purposes. McClelland concluded that the largest remaining challenge was an identification of *Ulluchu.*

Issues Surrounding the Name Ulluchu

The name *Ulluchu* seems to have been coined by Larco (1939). According to his description, the name originated in the Virú River valley, and is supposedly of Mochica origin. However, there is no linguistic evidence that such a term existed in the Mochica or Yunga language. The most comprehensive Mochica-Spanish dictionary available, compiled from the writings of Moche scholar E. Brüning (2004), has no such term. In addition, the local population - as well as market vendors, plant collectors, and *curanderos* interviewed - had no knowledge of *Ulluchu* whatsoever, other than what they derived from Larco. Since this first publication, the term has been copied by all subsequent authors (McClelland 1977, Wassen 1987), without any regard to its validity. It is unlikely that Brüning (2004) would have missed the name when doing his research early in the 20th century, if it indeed was still being used. Brüning lists quite a few Mochica plant names, some of which are still used for the same plants today, e.g., "*faik*" = *Acacia macracantha* (*faique, espino*), from "*fáçek, fáçke*" = spine.

The only language that has a somewhat similar word from which *Ulluchu* could be derived is Quechua: "*uchu*" translates as "chili, pepper," while "*ullu*" translates as "penis." The term "*ullu uchu*" is sometimes used as a name for *Columellia ovata* R. & P. (Columelliaceae), a small high-altitude plant described as "a very thick tree; its wood is suitable for various purposes, and its leaves have febrifugal properties and are very bitter" (Ruiz 1777). However, this plant has no likeness whatsoever to the Moche *Ulluchu*. Thus we must conclude that the term *Ulluchu* was most likely coined by Larco (1940: 98) based on a Quechua term for a species with somewhat similar fruits that has no relation to the species used by the Moche.

Botanical Identification: Why is Ulluchu not Carica candicans?

Carica candicans is a wild relative of papaya (*Carica papaya* L.). Although the fruits are not marketed, they are occasionally consumed by the local population and some market vendors and healers interviewed did know the plant under its vernacular name, "*mito.*" Larco (1939; 1940) never mentions the plant in relation to *Ulluchu*. Assuming that he indeed encountered a plant with that name, it cannot have been *C. candicans*, because this species would have been named "*mito.*"

McClelland (2004) argues that *Ulluchu* "cannot be a papaya, which belongs to a group of plants called 'cauliflory' [i.e., stem flowery]. . . . The flowers and fruit of a cauliflory grow on the trunk of the tree and not on the limbs . . . . *Ulluchus* depicted in Moche art, however, hang from limbs. Papaya leaves do not resemble *ulluchu* leaves, which are triangular, ovoid, or boomerang shapes hanging from limbs. Each large palmate papaya leaf grows on a stem from the top of the tree. I However, further complicating this matter, it turns out that *C. candicans* happens to be one of the few papayas that are not cauliflorous, that have triangular leaves with entire margins, and the fruits do hang from branches. Thus, judging from the iconography alone, *C. candicans* actually could be *Ulluchu*.

However, based on accumulated archaeological evidence (Alva et al. 1993, Alva 1994, Donnan et al. 1994, McClelland 2008), it has become clear that the actual fruits found in burials do not resemble *C. candicans* (Fig. 11). In addition, the explanation that papain might have

Figure 11: Front view of crushed Ulluchu. Photo: Donald McClelland.

been extracted by the Moche from unripe papaya for use as a blood anti-coagulant (Hultin et al. 1987), albeit reasonable, does not make much sense from a phytochemical perspective. Cultivated papaya (*C. papaya*) is often depicted in Moche pottery and the species contains large amounts of papain. Why would the Moche have resorted to a rare wild species, when they could have used a cultivar with the same properties that grew on their very doorstep? Also, *C. candicans* fruits are often 10-15 cm long and, while this would relate to the size of some of the *Ulluchus* in the iconography, it is vastly larger than many of the fruits depicted and much larger than the fruits found in burials. Finally, the anatomy of *C. candicans* simply does not correlate with the fruits encountered in burials.

What is Ulluchu, and what was it really used for?

Moche fine-line drawings of *Ulluchu* normally depict seed pods or seeds floating in the air in sacrificial scenes (McClelland 2008: Fig. 3.14), associated with runners and messengers (e.g., McClelland 2008: Fig. 3.1) or intoxicated priests (e.g., McClelland 2008: Fig. 3.6). The *Ulluchu* fruits vary greatly in size, ranging from about 1-15 cm. They are normally comma-shaped, often with an "exaggerated round calyx" (McClelland 2008: 43) with lines on the body of the fruit (e.g. McClelland 2008: Fig. 3.4). Some illustrations show *Ulluchu* harvested by monkeys. In these cases the fruit is mostly shown growing along the axes of the plant's leaves (e.g., McClelland 2008: Figs. 3.27 & 3.28).

Starting from this basis in 2002, we built on the work of Donna McClelland and the archaeological excavations at Sipán in the Lambayeque Valley (McClelland 2008; Alva 1994; Alva et al. 1994) and at Dos Cabezas in the Jequetepeque Valley in the 1990s (McClelland 2008; Donnan 1994). Botanically, all these depictions resemble capsules or drupe-like fruits. It became apparent that in a biodiversity hotspot like Peru, with a flora of more than 18,000 species, a large number of plant families have fruits that vaguely resemble Moche fineline drawings of *Ulluchu* and many of these families contain more than one genus with similar fruits. Examples include: (Apocynaceae: *Ambelania*; Caricaceae: *Carica*; Celastraceae: *Maytenus*; Chrysobalanaceae: *Chrysobalanus, Hirtella, Licania*; Convolvulaceae: *Dicranostyles*; Fabaceae: *Aldina, Alexa, Andira, Dipteryx, Dussia, Ormosia*; Guttiferae: *Tovomita*; Hippocrateaceae: *Cheiloclinum, Salacia*; Icacinaceae: *Calatola*; Meliaceae: *Guarea;* Menispermaceae: *Abuta, Curarea, Elephantomene, Telitoxicum*; Myristicaceae: *Virola*; Olacaceae: *Cathedra*; Quinaceae: *Lacunaria*; Sabiaceae: *Meliosma*; Sapindaceae: *Cupania, Paullinia* and Sapotaceae: *Pouteria*. Some of these are still highly important in traditional societies. For example, *Ambelania* fruit is often consumed. *Ormosia* contains potent poisonous compounds, but is now mostly used in crafts; *Curarea* is one of the ingredients of "*curare,*" the famous Amazonian arrow poison. *Virola* species are still used as potent snuffs in the Amazon. Finally, *Paullinia* is the source of "*yopo,*" an important stimulant. However, none of these carry the vernacular name *Ulluchu*.

Fortunately, at this tine the archaeological evidence provides good clues for identification. The *Ulluchu* fruits found in burials in the 1990's are clearly capsules or drupes, slightly comma-shaped, between 1.5 and 5 cm long and slightly grooved. They closely resemble bone, gold and *Spondylus* beads found *in situ* in size, form, and texture. In the iconography, the fruits are often depicted as located on both sides of branches drawn on headdresses. It is important to note that the person wearing the headdress has widely extended nostrils as is often seen in people inhaling hallucinogenic snuffs. He is also holding a gourd and pestle. McClelland (2008: Fig. 3.18) interpreted this as a lime gourd used for chewing coca. We suggest that this might also be a gourd used to grind the seeds of *Ulluchu* for inhalation. Further iconographic evidence supports this hypothesis. Runners and messengers associated with *Ulluchu* are often winged. They literally fly, i.e., the inhalation of *Ulluchu* gives them wings. The personages have *Ulluchu* depicted on their belts, *Ulluchu* seeds floating above their heads, and they hold instruments that closely resemble a double snuff tube that would serve to inhale powdered hallucinogenic substances. Thus, it seems possible that one of the uses of *Ulluchu* may have been as a mind-altering snuff. Another reason for identifying the seeds in the iconography as hallucinogenic *Ulluchus* is underscored in Moche paintings, e.g., a famous Moche scene (McClelland 2008: Fig. 3.34) where monkeys are picking fruits from an *Ulluchu* tree. It is important to note that the tree depicted has opposite leaves and that seeds are extracted from the fruit, possibly for roasting in an oven on the bottom right. The roasted seeds could then be ground into powder and inhaled. The fruits themselves seem to be 5-valved. The function

Figure 12: Line of prisoners at Huaca El Brujo. Photo: Rainer W. Bussmann

of *Ulluchu* as a hallucinogen is reinforced by other imagery (e.g., McClelland 2008: Fig. 3.6) where personages surrounded by *Ulluchu* fruits lie on the ground in what appears to be an intoxicated state. In addition, prisoners in sacrificial scenes (e.g., Hocquenghem 2008: Figs. 2.2, 2.3, 2.12 & 2.24), especially the well known "lines of prisoners" at Huaca El Brujo and Huaca de la Luna, all show clearly visible erections (Fig. 12), which may be seen as another indication of the ingestion of a substance which would cause such an effect. From this perspective Larco's term *Ulluchu* (Larco 1940), if derived from Quechua "*ullu-uchu*" or "penis pepper," would in fact make sense in describing the possible effects of the plant in question. Also, the association with sexual arousal is reinforced by a mythical scene where an *Ulluchu* tree is in the background of a scene of ritual copulation.

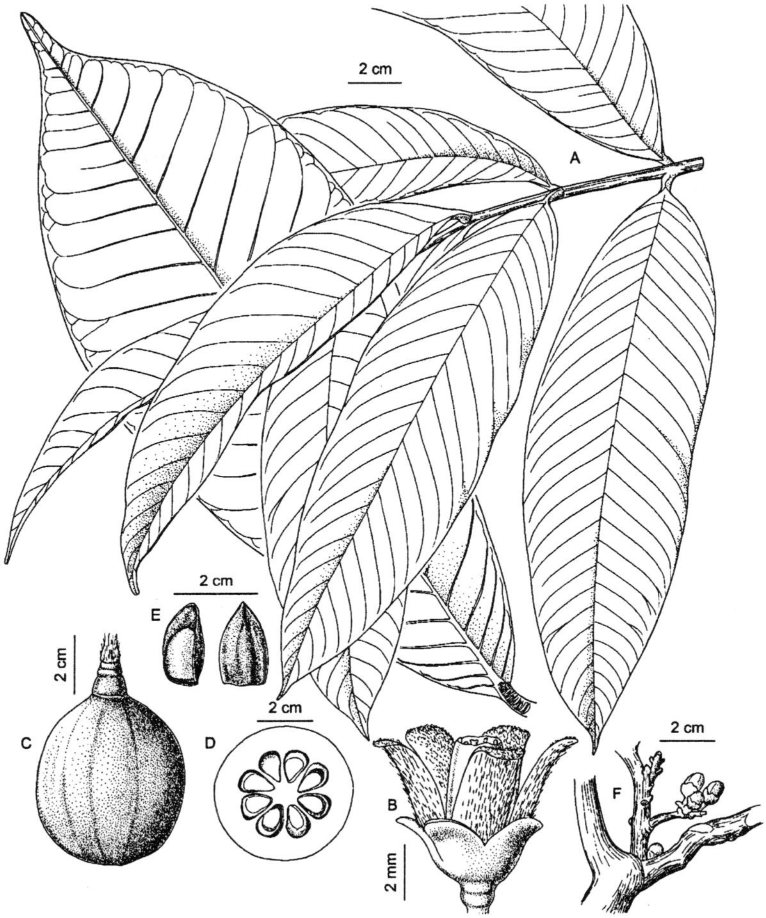

Figure 13: Guarea grandifolia
A. Mature branch, B. Flower, C. Mature fruit, D. Fruit cross section, E. Seeds, F. Branching pattern

In light of the above, *Ullucho* is a tree with opposite leaves and fruits that are drupes between 1-15 cm long possibly containing psychoactive ingredients that would elevate the blood pressure and cause erections. The only plant family from the list above having representatives that meet all these criteria is Meliaceae, and the genus *Guarea* is the one that most closely fits the description. It includes trees with pinnate leaves (which is unusual for Meliaceae) and fruits that are 3-5 valved capsules with large, pseudo arillate seeds. The genus *Guarea* is found throughout Peru, but is mostly restricted to tropical lowland forests, with some species reaching cloud forest habitat. No species is found along the dry coast of Peru, which indicates that the material must have been widely traded in Moche times. A typical representative is *Guarea grandifolia* DC. (Fig.13). The species has clearly pinnate leaves, and the fruits very clearly resemble the archaeological samples. In addition, *Guarea* contains a large number of species with varying fruit sizes (from 1-15 cm), swollen calyx, and grooving on the body of the fruit, which all correlates with the *Ulluchu* imagery in Moche fineline drawings. The seeds of *Guarea* species, with a distinct white navel, very much resemble the seeds depicted in Moche fineline paintings.

Many species of *Guarea* are known by a wide variety of vernacular names, e.g., *Guarea* spec.: requia, kushímsakish; *G. glabra:* yecheñor, yechemor; *G. grandifolia:* bola requia; *G. guidonia:* atapio, latapi, latapi caspi, requia colorada, requia latapi, xoro; *G. kunthiana:* requia, paujil ruro; *G. purusana:* latapi, requia). The wood of many species is used as timber for construction. The wood, bark and leaves contain compounds that act as abortive, emetic, purgative, and antiamoebiac agents and the bark is often sold as *Coccilliana* in expectorant preparations (Kraemer 1915; Rätsch 1998). The fruits and seeds contain a variety of alkaloids that are very rarely used due to their high toxicity (Kraemer 1915). Some of the alkaloids found, e.g., rusbyine, have a structure and effects like emetine, an alkaloid found in *Psychotria ipecacuanha* (Brot.) Stokes, which has been widely used as an emetic and expectorant. Other species of *Psychotria* are well known as components in *ayahuasca* preparations due to their high content of N,N-DMT (Rätsch 1998). In large doses, *ipecac* preparations cause high blood pressure, arrhythmia, spasms, and extension of the blood vessels. While the existing literature on *Guarea* seed compounds is rather fragmentary, it seems clear that a concentrated dosage of *Ulluchu* seeds, if ingested, would increase a prisoner's heartbeat, elevate the blood pressure, and widen blood vessels thus causing erection. All of this would make it much easier to extract sacrificial blood. Also, when inhaled by priests, the active compounds could have a mind-altering effect, which would not necessarily lead to high levels of toxicity, and could induce very rapid, short-term hallucinations.

We conclude that *Ulluchu* can be identified as a group of species of the genus *Guarea* (Meliaceae) based on morphological characteristics. In addition, the chemical composition of the plant's compounds supports the thesis that it was used in a sacrificial context to improve the extraction of blood from sacrificial victims. We also suggest that a ground preparation of *Guarea* seeds, when inhaled, may have been used as a hallucinogen. However, more detailed phytochemical research is needed to corroborate the latter hypothesis.

Changing Markets

Exotics played an important role among plants sold in Norhern Peruvian markets. Fifty-nine species (15%) found in all markets were exotics. However, among the species most commonly encountered in the inventories, 40-50% were exotics. *Matricaria recutita* (chamomile) was found in the inventory of approximately 70% of vendors. The next most popular species sold in these markets included *Equisetum giganteum, Phyllanthus urinaria, Phyllanthus stipulatus, Phyllanthus niruri* (Chanca Piedra or Stone Breaker), *Eucalyptus globulus (eucalyptus), Piper aduncum, Uncaria tomentosa* (Cat's Claw), *Rosmarinus officinalis* (Rosemary), *Peumus boldus, Bixa orellana* (Achiote), and *Buddleja utilis.* However, when taking sales volume into account, *Croton lechleri* (Dragon's Blood), *Uncaria tomentosa,* and *Eucalyptus globulus* were clearly the most important species (Bussmann and Sharon 2009c).

While it was very easy for all vendors to name their most important and frequently sold species, it proved impossible to get detailed information about species that vendors considered to be "rare" or "disappearing." In most cases, vendors mentioned species as rare because they themselves did not sell them, i.e., in such cases these plants were very common outside the market (e.g., *Plantago major,* common plantain) or, because demand was so low, it made no sense to carry them in their inventories.Very small vendors had inventories that represented the most common medicinal plants available and excluded most species in the large "witchcraft" or "sorcery"segment of the pharmacopoeia. On the other hand, well-established large stands specialized in supplies for healers, including "magical" plants, amulets, perfumes, etc.

All four markets had inventories containing more than 50% of all inventoried plant species, but lacked many of the "generalist" plants sold by smaller vendors. The portfolio of these stands focused almost entirely on "magical" species that are needed to cure illnesses like "*susto*" (fright), "*mal aire*" (evil wind), "*daño*" (sorcery), "*envidia*" (envy) and other "magical" or psychosomatic ailments. At the same time, to meet tourist demands all four markets catered to the esoteric tourism market that tends to frequent the large markets which carry a variety of plants not used by *curanderos*.

A look at sustainability - How much plant for what price?

More than two-thirds of all species sold in Northern Peruvian were alleged to originate from the highlands (*sierra*), i.e., the intermontane valleys which are often heavily used for agriculture and livestock grazing. The overall value of medicinal plants in these markets is US $ 1.2 million per year. This figure only represents the share earmarked by market vendors and does not include the amount local healers charge for their services. Thus, medicinal plants contribute significantly to the local economy. Such a large market raises questions of the sustainability of this trade, especially since the market analysis does not take into account any informal sales.

Most striking was the fact that seven indigenous and three exotic species, i.e., 2.5% of all species traded, accounted for more than 40% of the total sales volume. Moreover, 31 native species accounted for 50% of all sales, while only 16 introduced plants contributed to more than a quarter of all material sold. This means that a little over 11% of all plants in the market accounted for about three-fourths of all sales. About one third of this sales volume includes all exotic species traded. None of these are rare or endangered. However, the rising market demand might lead to increased production of these exotics, which in turn could have negative effects on the local flora (Bussmann and Sharon 2007b).

A look at the indigenous species traded highlights important conservation threats. *Croton lechleri* (Dragon's Blood) and *Uncaria tomentosa* (Cat's Claw) are immensely popular at a local level and each contributes to about 7% of the overall market value. Both species are also widely traded internationally. The latex of *Croton* is harvested by debarking the whole tree. *Uncaria* is mostly traded as bark, and again the whole plant is normally debarked. *Croton* is a pioneer species and, apart from *C. lechleri*, a few other species of the genus have found their way into the market. Sustainable production of this genus seems possible, but the process has to be closely monitored. However, the current practice does not appear sustainable since most *Croton* is wild-harvested. In the case of Cat's Claw, trade is so large that years ago collectors of this primary forest liana started complaining about scarcity (Cabieses 2000). During the years of this study other *Uncaria* species, and even *Acacia* species, have appeared in the market as "Cat's Claw" (personal observation). Under such conditions the *Uncaria* trade is clearly not sustainable.

Some of the other "most important" species are either common weeds (e.g., *Desmodium molliculum*) or are part of large populations (e.g., *Equisetum giganteum*). However, a number of species are very vulnerable. *Tillandsia cacticola* grows in small areas of the coast as an epiphyte

(Downer 2006). The habitat, coastal dry forest and shrub, is heavily impacted by urbanization and mechanized agriculture, the latter worsened by the bio-fuel boom.

Gentianella alborosea, G. bicolor, G. graminea, Geranium ayavacense and *Laccopetalum giganteum* are all high altitude species with very limited distribution. Their large-scale collection is clearly unsustainable and, in the case of *Laccopetalum,* collectors indicate that supply is harder and harder to find. The fate of a number of species with similar habitat requirements raises comparable concerns. The only species under cultivation at this point are exotics and a few common indigenous species.

In order to determine if Peruvians have a preference for medicinal plants or pharmaceuticals and the reasons why, we conducted three surveys in clinics located in Trujillo: a conventional Western-style private clinic (Bussmann, Sharon, and Lopez 2007), a private herbal and homeopathic "laboratory" (Bussmann, Sharon, and Garcia 2009), and a social security center for complementary medicine (Fajardo and Sours 2013). Results indicate that the major reasons for preference are fairly obvious. Those people who prefered using plants more often did so because using plants that have been in use for centuries appeared to be a safer and healthier alternative. When interviewed, these people acknowledged that pharmaceuticals are used for particular illnesses, but often have side effects that result in negative impacts elsewhere in the body. However, most respondents agreed that pharmaceutical products appeared to be more effective than medicinal plants, at least in the short term. Even though they still used plants, they tended not to depend on them completely.

A lot of agreement was registered for the use of a doctor's prescriptions. Many people had faith in their doctor. If their physician recommended using a certain medicine, they complied. This faith was based on the confidence people have in scientific medicine which produces a great deal of research that is available to the public. In addition to the research, respondents knew that medicine has noticeable effects that can be more quickly obtained than those resulting from plant use. Plant remedies take longer and are more subtle in their effects. These reasons were given by those who used pharmaceuticals most often. Although the number was minimal, there were respondents who did say that they used the two kinds of medicine in the same amounts. What was interesting was that most respondents said that they used both medicines together. For example, respondents often said that they would drink a cup of herbal tea while taking pills. Although people felt that each type of medicine has a role, most agreed that pharmaceuticals provide the best route for treating many biomedical illnesses. An interesting fact that emerged from our study of complementary medicine was that the government's annual statistics for its 26 clinics measure "reduction or elimination of pharmaceuticals" as one of its benchmarks of successful therapeutic outcomes.

In spite of public trust of modern medicine, Traditional Medicine in Peru is experiencing increasing demand as indicated by the fact that the number of herb vendors, particularly on the peripheries of major markets, has increased in recent years. Also, a wide variety of medicinal plants from Northern Peru can be found in urban supermarkets as well as the global marketplace. While this trend might help to maintain traditional practices and to give traditional knowledge the respect it deserves, it poses a serious threat as indicated by signs of over-harvesting of important species.

Figure 14: Sustainable plant use from plant documentation with local healers to production of correctly identified bulk herbal packages and teas for local markets.

Figure 15: Dissemination of traditional knowledge in bilingual books.

Today the most serious threat to this millennial tradition is the destruction of medicinal plant habitats. Urban sprawl and the sugar industry have already greatly altered the coastal plains around Trujillo and Chiclayo. Climatic change and deforestation are threatening the mountain forest systems that are the source of many medicinal species. Also, the high Andean ecosystems and sacred lagoons where many medicinally active species are found are in danger of being destroyed by large-scale mining activities (Downer 2007, Zamora 2007). To counteract such influences, extensive work with healers in documenting local knowledge is required in order to apply the results in developing sustainably sourced and scientifically identified plant material for local markets. Fortunately, such work is currently being pursued (Fig. 14). The knowledge is also being repatriated in bilingual form (Fig. 15).

Overall the respondents to our three surveys used medicinal herbs more often than pharmaceutical medicines, but only to a small degree. They generally assumed that plants are healthier and safer to use because they are natural and are thought not to have any side effects. But it was difficult to determine from these studies if the knowledge of the use of medicinal plants was growing or decreasing since our surveys indicated that respondents believed that the last generation knew more than the present. Nonetheless, most of the current generation does teach their children about the use of medicinal plants. Our studies also showed what medicinal plants the respondents used for which purposes. Similarly, the plant knowledge of patients at all three facilities surveyed was largely identical, with an essentially overlapping selection of common, mostly introduced species and their uses. This indicates that traditional medicinal knowledge is a major part of the people's culture that is being maintained while at the same time patients are also embracing the benefits of Western medicine.

The situation described above does lead to profound challenges when it comes to the safety of the plants employed, in particular for applications that require long-term use. Bussmann et al. (2013) found that various species were often sold under the same common names. Some of the different fresh species were readily identifiable botanically, but neither the collectors nor the vendors make a direct distinction between species (Fig. 16). Complicating matters even more, material was often sold in finely powdered form or in liquid extracts, which makes the morphological identification of the species in the market impossible and greatly increases the risk for the buyer. The best way to ensure correct identification would be DNA bar-coding. However the necessary technical infrastructure is not available locally. The use of DNA bar-coding as a quality control tool to verify species composition of samples on a large scale would require careful sampling of every batch of plant material sold in the markets. The volatility of these markets makes this an impossible logistical task. Often the same or closely related species mentioned in the literature sell under a wide variety of common names. Thus, one species might be sold as "Hercampuri" in one market stand while selling under a different name at a neighboring stand. As expected, there is no consistency in the dosage of plants used, nor do vendors often have any idea of possible side effects.

In the course of an evaluation of the phyto-therapy treatment modality of EsSalud's Center for Complementary Medicine in Trujillo (Gaucksheim, Narváez, Pons, and Sharon 2013) we came upon a major innovation with regard to quality control. An integral part of the prescription of medicinal plants in the center's natural pharmacy is a set of five quality control standards that their suppliers are required to meet, including botanical identification of species,

Figure 16: Medicinal plant presentation in Mercado Aviación Lima. A. Packaged ground plant material, B. Traditional presentation, C & E. *Geranium sessiliflorum*, D. Unidentifiable *Geranium* fragment, F. *Cheilanthes bonariensis*, G. *Argyrochosma nivea*, H. *Gentianella thyrsoidea*, I. *Gentianella nitida*.

microbiological analysis, non-metallic contamination certification, sanitary packaging requirements, and documentation of cultural value.

Our studies indicate that plant use in Northern Peru, although based on a millennial tradition, has changed considerably, especially during the last decades. Even in cases of plant species used for very clearly circumscribed applications, patients run a considerable risk when purchasing their remedies in the local markets and the possible side effects can be serious. More regulation and stringent identification of the material sold in public markets and entering the global supply chain via Internet sales is needed to overcome these difficulties. In this regard, the quality control model provided by EsSalud's phyto-therapy treatment modality is a step in the right direction.

Sacred Seeds, the Nagoya Protocol, and Repatriation of Traditional Knowledge

Sacred Seeds is one of the programs of the William L. Brown Center at the Missouri Botanical Garden. Creating a global model of sustainable development through botanical research programs, Sacred Seeds works closely with communities in host countries to promote long-term success in finding practical solutions to conservation issues (Fig. 17). The program aims to establish in-situ collections of locally important plant species used in Traditional Medicine as food and medicines. Participating sanctuaries receive direction, feedback, guidance, and advice from the WLBC team, linking experiences in botany, anthropology, traditional healing, and conservation with the collective experience of all Sacred Seeds sanctuaries. Sacred Seeds strives to facilitate seed storage training, with the possibility of building a relationship with large seed storage facilities. It also provides training in plant taxonomy and publishing assistance, including templates for creating educational, scientifically valid literature and a Sacred Seeds Book for in the local language, but also for worldwide distribution (Fig. 18). The Sacred Seeds program currently has almost 40 primary gardens, with a network of close to 1,500 participating garden units.

The Nagoya Protocol and Repatriation of Traditional Knowledge

The final approval of the "Nagoya Protocol on Access to Genetic Resources and the Fair and Equitable Sharing of Benefits Arising from Utilization of the Convention on Biological Diversity" has provided a great boost for the rights of indigenous and local communities.

The main objective of the protocol is "*the fair and equitable sharing of the benefits arising from the utilization of genetic resources, including appropriate access to genetic resources and appropriate transfer of relevant technologies, taking into account all rights over those resources and technologies, and appropriate funding, thereby contributing to the conservation of biological diversity and the sustainable use of its components.*" A major requirement is that "*traditional knowledge associated with genetic resources held by indigenous and local communities is accessed with prior informed consent, approval, and involvement of indigenous and local communities, and that mutually agreedupon terms have been established.*"

The latter requirement is of great importance. Fortunately, the establishment of prior informed consent has already been widely practiced by most major scientific institutions. However, there are still projects that place limited emphasis on permits and consent, since the process is often long and tedious. Under the Nagoya Protocol prior informed consent and provision of benefits to knowledge holders are no longer just a matter of good ethics, This it is also a matter of international law. It is our hope that all major granting agencies, whether private or governmental, will soon make proof of prior informed consent a requirement for funding.

The concept that "benefits" might result from the documentation of traditional knowledge is somewhat new to many colleagues. In globalized science, where data is easily circulated, it is only fair to assure that the knowledge our counterparts share with us is not appropriated by parties not involved in the original study, whether for scientific or commercial purposes. In practice, this means that the establishment of prior informed consent, as valited under the

Figure 17: Sacred Seeds Ethnobotanical Garden Network website

Figure 18: Sacred Seeds Garden Guide

Nagoya Protocol, needs to include an explicit statement along the lines that "*any work conducted in a community is carried out under the Nagoya Protocol on Access to Genetic Resources and the Fair and Equitable Sharing of Benefits Arising from their Utilization and that the right to use and authorship of any traditional knowledge by all informants is maintained. Also, any use of the information obtained, other than for scientific publication, requires additional prior consent of the traditional owners as well as a consensus on access to benefits resulting from subsequent use.*"

At the William L. Brown Center (WLBC) Missouri Botanical Garden (MBG), the Nagoya Protocol is immediately applicable. It not only covers future research, but any previous data gathered by our institution, unless a different agreement with the original owners exists.

Benefit sharing in this context needs to include not only the repatriation of new data gathered in a language and form accessible to the traditional owners, but also the translation and repatriation of the results of previous studies conducted in the same indigenous or local community, if not already done by the original researchers. In addition, should they so desire, informants must be allowed full participation as authors in all publications, rather than simply being mentioned in the acknowledgments.

In this regard WLBC has edited a series of books authored by the members of local communities, repatriating knowledge on plant use to the original knowledge holders, published in the local languages and in appropriate formats as requested by the communities. These books are given to all participating community members as well as to local schools. The first reaction in most communities when presented with the results of such a collaborative investigation is astonishment that the researchers actually returned and in fact brought useful material published in the local language. This is normally followed by great satisfaction on the part of the participants at being actual authors of the material (Fig. 19). All publications acknowledge the local community's intellectual property under the Nagoya Protocol (Fig. 20) as well as the authorship of the local collaborators (Fig. 21). In the case of publications on palm use, in cooperation with the communities WLBC developed easy to understand symbols for usage categories and plant parts used (Fig. 22). The books include graphic descriptions of plant parts (Fig. 23), with color examples for different growth, leaf, and fruit-forms (Figs. 24 and 25), followed by an in-depth description of the individual species (Fig. 26) and detailed, well illustrated desriptions of usage (Fig. 27). In addition to these efforts, WLBC is striving to translate previously published studies on traditional knowledge making them available for the benefit of the communities involved in the original research. A good example is Bussmann and Paniagua Zambrana (2011), "La etnobotánica de los Chacobo," a translation of "The Ethnobotany of the Chacobo Indians, Beni, Bolivia" by B. Boom (Fig. 28).

Figure 19: Repatriation of knowledge – Presentation of palm books to local communities

"KAMPANAK SE USA PARA EL TECHO PERO YA NO HAY"
USO Y CONSERVACIÓN DE PALMERAS ENTRE LOS AWAJUN, AMAZONAS, PERÚ

Este libro forma parte del material generado por el Proyecto Impacto de la cosecha de palmeras en los bosques tropicales (PALMS), financiado por el 7º Programa Marco de la Unión Europea, contrato nº 212631 (http://www.fp7-palms.org) y las instituciones colaborantes.

Herbario Nacional de Bolivia
Instituto de Ecología – Universidad Mayor de San Andrés (UMSA)
Campus Universitario, Calle 27 Cota Cota s/n
Casilla 10077 correo central
La Paz, Bolivia

Departamento de Biología - Área de Botánica
Universidad Autónoma de Madrid (UAM)
Edificio de Ciencias Biológicas

Impreso en GRAFICART SRL:
San Martín 375 - Trujillo, Perú
ISBN-10: 0984841555
ISBN-13: 978-0-9848415-5-4
Hecho el Depósito Legal en la Biblioteca Nacional del Perú N° 2012-07672
Primera edición, Trujillo, Perú, julio del 2012

Todos los trabajos se ejecutaron bajo el reglamento del "Protocolo de Nagoya sobre acceso a los recursos genéticos y participación justa y equitativa en los beneficios que se deriven de su utilización al Convenio sobre la Diversidad Biológica" y se mantiene el derecho de autoría de sus conocimientos tradicionales de todos los informantes. Cualquier uso comercial de la información presentada en esta publicación requiere del consenso previo con los informantes y comunidades, y un acuerdo sobre la distribución de beneficios.

IMPRESO EN PERÚ - PRINTED IN PERU

Figure 20: Book page acknowledging local intellectual property and the Nagoya Protocol

AUTORES POR COMUNIDAD

PALMA REAL

Adela Shanocua
Aidé Chetu
Alejandro Mohahe
Alfonsa Sehua
Alfredo Biaeja
Alicia Saavedra
Alina Shoe
Ana Elinei
Asunta Zapata
Benito Huane
Betti Biaeja
Carlitos Tirina
Carlitos Tirina
Carmen Machuquia
Cecilio Yoyaje
Celestina Hejayha
Zenón Yohahe
Clementino Meshi
Dali Tipuna
Dorilda Posho
Doris Elinei
Elda Hudenda
Elisa Huajohuaho
Eliseo Meshi
Eliser Biaeja
Elvira Saavedra
Esperanza Heyahijia
Exilda Shanoqua
Felipe Vauhehua
Félix Huali
Florentino Montesinos
German Biaeja
German Biaeja
Gloria Saavedra
Graciela Huohehua
Gregorio Mejía
Héctor Ioshi
Henry Posho
Inés Meshi
Asunta Biaeja
Jesús Meshi
Jesusa Hudomela
Jordán Kioshi
Jorge Shanoqua
José Gongora
José Liaqui
Geovana Chaeta
Juan Shanoqua
Juanita Ekinei
Julia Huane
Lidia Meshi
Lidia Shehua
Lorenzo Shoe
Lucio Yohahe
Luis Biaeja
Luzmila Shanokua
Marcela Biaeja
Marcial Soria
Mardilina Biaeja
Margarita Meshi
María Biaeja
María Isabel Saavedra
María Naji
Maricruz Shanoco
Mateo Biaeja
Melchor Vauhehua
Mercedes Shanocua
Mercedes Shehua
Midio Huane
Miguel Domínguez
Nina Saavedra
Pastora Huaquibeline
Pilar Valdivia
Pio Meshi
Raúl Meshi
Ricardo Yoyaje
Roger Mahio
Rosa Paketanaji
Rosario Soria
Rosy Viasteno
Rudi Sehua
Ruth Tilihua
Sara Saavedra
Segundo Saavedra
Teófila Saavedra
Verónica Beshu
Victoria Meshi
Walter Ayca
Wilma Huahojehua
Yasira Shehua

Figure 21: Community authors

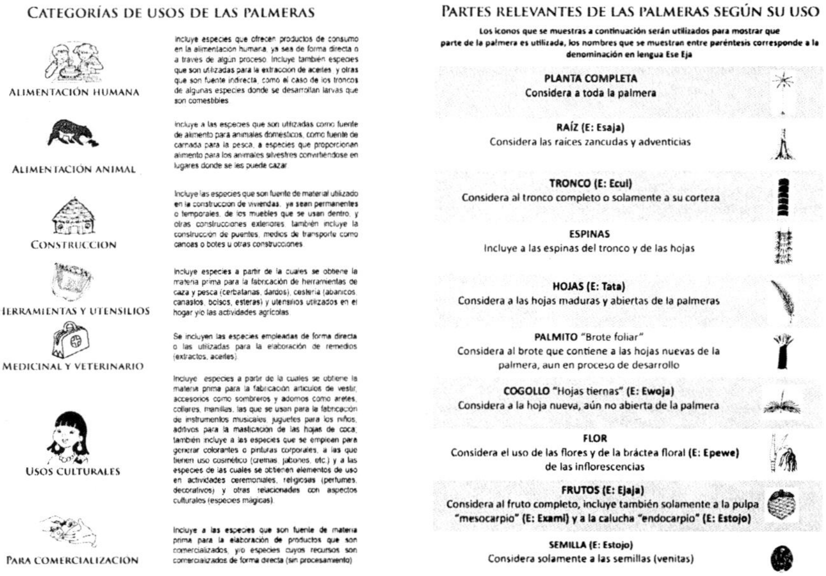

Figure 22: Symbols to explain plant uses and plant parts developed in co-operation with local communities

La estructura de las palmeras

Para entender la diversidad que existe entre las palmeras que conocen las comunidades Llacuash de la región de San Martín, es necesario familiarizarse con la estructura de la planta y los nombres correspondientes. Para ello presentamos en esta sección una vista rápida de la estructura de las palmeras.

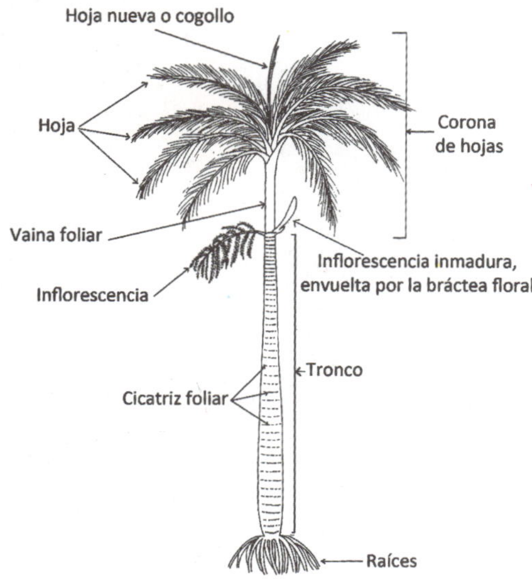

Figure 23: Palm profile – plant parts in local language

Figure 24: Examples for different palm leaf and growth forms

Figure 25: Examples for palm fruits

ESHISHI

Nombre castellano: Shapaja

Nombre científico: *Attalea phalerata* Mart. ex Spreng.

DESCRIPCIÓN DE LA PLANTA:

Palmera: Tallo solitario de hasta 14 m de alto; 25 a 40 (60) cm de diámetro; tronco densamente cubierto con las bases persistentes de las hojas.

Hojas: De 11 a 30 aproximadamente de 2 m de largo, con 245 pinnas irregularmente dispuestas a cada lado e insertadas en diferentes planos que le dan a las hojas una apariencia desordenada.

Flores: Con los dos tipos, masculinas y femeninas, en una misma planta (monoica). Tanto las masculinas como las femeninas en racimos (inflorescencias) que salen del medio de la corona de hojas (interfoliares), con un pedúnculo largo, péndulas colgando a los lados de la palmera. Flores femeninas globosas, amarillentas y masculinas numerosas, pequeñas, amarillentas las masculinas, y con un fuerte olor dulzón.

Frutos: Ovoides, 6 - 11 cm de largo, con una cáscara dura (exocarpio), amarillenta cuando madura; pulpa (mesocarpio) interna, aceitosa, suave, cremosa, de color amarillo hasta anaranjado intenso, dulce.

Semillas: Dentro un hueso leñoso (endocarpio o calucha) y fibroso, 1 a 7 por fruto, alargadas, similares a almendras.

USOS:

En los troncos caídos se desarrollan las larvas de un escarabajo, llamadas **suri (E: Sooso)**, que son recolectadas para comerlas fritas.

En algunos casos se recolecta el **suri (E: Sooso)**, para emplearlo como carnada para la pesca.

Algunas veces recolectan el **suri (E: Sooso)**, y extraen su aceite friéndolos; este aceite filtrado se toma por cucharadas con miel y limón para curar la tos fuerte, neumonía y bronquitis.

Las hojas son utilizadas para la construcción de techos en las casas. Previamente al techado y secado son tratadas de dos formas diferentes: la primera, consiste en partir las hojas por la mitad, al nivel de la nervadura central, las que luego de unen y secan juntas, esta técnica es la de las hojas partidas. La segunda, consiste en realizar un corte superficial a nivel de la nervadura central, lo que permite doblar las pinnas de un lado hacia el otro, de tal forma que quedan en una sola fila, esta técnica es la de las hojas rayadas.

Las hojas dobladas ocasionalmente son usadas para construir **cercos** (paredes) temporales en las casas.

El uso mas frecuente de las hojas es para tejer el **surubí** o **cumba** que se colocan en la parte superior de los techos.

Las hojas tiernas o **cogollos**, se pueden tejer y sirven para confeccionar abanicos **(E: Epeji)**, o **ventadores**, canastos **(E: Esaja)**, esteras **(E: Eadiji)** y sombreros **(E: Ehaoha).** Antiguamente también se las usaba para tejer canastos temporales usados cuando salían a cazar para traer la carne y cargas del bosque.

Cuando se corta un palma, se aprovecha para sacar el **palmito** (brote de hojas tiernas) que se come crudo, en ensalada, o cocido.

La **tola** (bráctea floral) es recolectada y utilizada ocasionalmente como recipiente (tipo batea) en la cocina.

Figure 26: Illustrated plant description

Figure 27: Illustrated description of plant use, including local language terms and Spanish translation

Figure 28: Repatriation of local knowledge in local language: Translation of Boom's "Ethnobotany of the Chacobo Indians"

A Global Program for Conservation of Useful Plants and Traditional Knowledge: A Call to Action

On 1–2 May 2013 WLBC brought international experts on plants used by humankind together in St Louis, Missouri to consider the ways in which a global crisis now underway - the loss of tens of thousands of plant species - could be addressed. These threatened plants include species vital to the lives of people throughout the world, including plants used for food and nutrition, medicine, cultural and spiritual purposes, and the maintenance of livelihoods. They are needed to redress poverty, provide food security, and ensure sustainable development in many nations. Plants and their associated biocultural knowledge play an essential role in the ecosystems that support all life on Earth (Barve et al. 2013).

The resulting statement is not only an appeal to the international community to address the tragic loss of plant diversity but a call for the development of a concerted effort worldwide to address the loss of essential knowledge about plants and their uses, especially at the level of local communities.

The meeting specifically focused on the objectives of the Global Strategy for Plant Conservation (GSPC), an initiative adopted by the U.N. Convention on Biological Diversity in 2002, and subsequently updated in 2010, as well as the GSPC targets pertaining to the maintenance and preservation of useful and culturally significant plants. The participants concluded that there is also great urgency in addressing the vital importance of traditional knowledge about plants, their utility, management, and conservation. This unique, ancient, and detailed knowledge is typically maintained by local and indigenous communities.

The workshop contributors urged the development of a global program on the conservation of useful plants and associated knowledge, taking into account the need to:

- Call on the international community and governments to recognize the importance of wild and cultivated plant diversity as well as the associated knowledge of its usefulness as a vital present-day and future resource. This should be accomplished through the successful implementation by 2020 of the GSPC targets and objectives.

- Highlight the need for a concerted international effort to compile a widely accessible global catalogue of useful plants of importance to humankind while respecting intellectual property rights, local ownership of knowledge, and appropriate benefit sharing.

- Assist local peoples in the preservation of their traditional knowledge in a culturally appropriate manner.

- Stress the need for cross cultural and multilevel partnerships in an effort to build on and share experiences in the conservation of culturally significant plants, their sustainable use, and associated knowledge.

- Develop an international research platform to address gaps in scientific knowledge of useful plants.

- Facilitate capacity building and training opportunities in ethnobotany, particularly in countries and regions with significant gaps in such resources.

- Support and encourage biocultural knowledge transmission and custodianship.

- Develop the appropriate facilities, methodologies, and techniques to support culturally sensitive curation of biocultural collections (artifacts, herbarium vouchers, products, living collections, etc.) and associated traditional knowledge.

- Elaborate and disseminate educational materials and resources in appropriate languages that support and promote the study and use of traditional knowledge and insure its inclusion in educational curricula.

- Develop a toolbox of methodologies, case studies, manuals, and best practices in order to support the conservation of useful plants and associated knowledge.

- Highlight the need for measurable indicators that monitor progress in the conservation of useful plants and associated knowledge.

- Follow the framework provided by the Nagoya Protocol on Access to Genetic Resources and the Fair and Equitable Sharing of Benefits Arising from their Utilization (ABS) as a supplement to the Convention on Biological Diversity and FAO's International Treaty on Plant Genetic Resources for Food and Agriculture in order to achieve ethical standards for access, fair and equitable benefit sharing, traditional resource and farmers' rights, and the protection of intellectual property.

Conclusions

Northern Peru occupies the upper sector of the "health axis" of the Central Andean cultural area. Few other places on the planet boast a medicinal flora as rich as this region. Archaeological evidence traces the use of plants used in traditional healing and divination practices of the region back at least 2,500 years. Current research shows that the composition of the local pharmacopoeia has changed since colonial times (Sharon and Bussmann 2006) and that the overall number of medicinal plants employed seems to have increased. This indicates that the North Peruvian health tradition is still going strong, and that the healers are constantly experimenting with new remedies. Recent indications of this are the appearance of Noni (*Morinda citrifolia*) products in large quantities in plant pharmacies and markets in the region. This makes it quite obvious that local herbalists are carefully watching international health trends to include promising species in their own repertoire.

The use of sacred psychoactive plants (entheogens), in particular the San Pedro cactus (*Echinopsis pachanoi*) is a vital component in Andean healing practices, and has been around for millennia (Camino 1992; Polia 1998; Sharon 1978; Sharon 1994; Joralemon and Sharon 1993). Archaeological evidence for this sacred cactus has been found at Guitarrero Cave (8200-6800 B.C.) in the sierra (Lynch 1980: 102), Las Aldas (2000-1500 B.C.) on the north-central coast (Fung 1969: 43, 120, 195), and Garagay (1643-879 B.C.) on the central coast (Burger 1992: 63-64). San Pedro is frequently depicted in Cupisnique, Chavín, Moche, Nasca, and Lambayeque iconography (Polia 2000; Sharon 2000; Glass-Coffin, Sharon, and Uceda 2004; Franco 2009). At Chavín, Torres (2008) has also identified *Anadenathera* (*vilca, cebil*), *Brugmansia (borrachero, floripondio, misha), Nicotiana (tabaco), and Erythroxylum (coca)* in the religious iconography of the site. Five hundred years of suppression of traditional healing practices starting in colonial times and continuing to manifest in the prejudices of contemporary national administrations have not managed to destroy this tradition. The use of San Pedro, together with additives like Angel's Trumpet (*Brugmansia* spp.) and tobacco (*Nicotiana tabacum* and *Nicotiana rústica*), is still a central part of the curing ceremonies of healers in Northern Peru. Healers are in fact experimenting with what for them are new hallucinogens, with some northern *curanderos* including decoctions of *Ayahuasca* (*Banisteriopsis caapi*) in their rituals.

Healing altars (*mesas*) in Northern Peru often follow the old tradition by including all kinds of "power objects," frequently with a "pagan" background. Objects such as seashells, pre-Hispanic ceramics, staffs, stones, etc. are very common on Peruvian *mesas*, and are blended with Christian symbols such as crosses and images of saints. Patients are cleansed by spraying them with holy water and perfumes, and herbal baths or "spiritual flowerings" (*baños de florecimiento*) are very important components of the healing tradition. In most cases the cleansing of the patients involves the nasal ingestion of tobacco juice and perfumes. While the incantations and songs *(tarjos)* used by healers during their curing sessions include Christian components, e.g., the invocation of Christ, the Virgin Mary and any number of saints, references to Andean cosmology, e.g., to the sacred lagoons (*lagunas)* and mountain spirits (*apus)*, are very common as well. The use of guinea pigs as diagnostic instruments is standard.

The knowledge of medicinal plants is still taught orally, with no written record. An illustrated identification guide for the medicinal plants of Northern Peru and their uses, similar to our field guide for Southern Ecuador (Béjar, Bussmann, Roa, and Sharon 2001; Bussmann and Sharon 2006a, 2007a) will hopefully help to keep the extensive traditional knowledge of this area alive. However, Traditional Medicine is experiencing increasing demand, as indicated by the fact that the number of herb vendors, an particular in the periferies of the markets of Trujillo, has increased in recent years. Also a wide variety of medicinal plants from Northern Peru can be found in the global market. While this trend might help to maintain traditional practices and to guarantee traditional knowledge the respect it deserves, it poses a serious threat, as signs of overharvesting of important species in the wild are becoming increasingly apparent.

Today the most serious threat to this millennial tradition is the destruction of medicinal plant habitats. Urban sprawl and the sugar, cotton, and rice agri-businesses have already greatly altered the coastal plains around Trujillo, Chiclayo, and Piura. Climate change and deforestation are threatening the sierra and mountain forest systems that are the source of many medicinal species. Most importantly, the high Andean ecosystems and sacred lagoons where many medicinal plants are found are in danger of being destroyed by large-scale mining (Downer 2006; Zamora 2007).

Finally, one threat to Traditional Medicine – prejudice - seems to be abating as evidenced by the fact that the Colegio Médico del Perú - which has resisted rapprochement with Traditional Medicine (TM) for decades - is now hosting workshops and colloquia in Lima dealing with this important topic in cooperation with EsSalud´s Dr. Martha Villar. In Trujillo, earlier we noted the proactive stance of EsSalud regarding the prescription of medicinal plants in their Center for Attention in Complementary Medicine (CAEM), and a volume on Traditional

Medicine published in 2009 by the Universidad Nacional de Trujillo - the latter partially based on a symposium ("Medicina y Cultura") organized in 1999 by Trujillo anthropologist Rafael Vásquez. And, it is especially noteworthy that antecedents for the latter anthropological work were developed in 1994 and 1995 by local archaeologists in two outdoor grassroots symposia ("Chacmas") including *curanderos,* local authorities, academicians, and the general public, which, unfortunately were not published—although the 1994 event was reviewed by Segura and Miranda (1995) and Morales (2012). As this book goes to press, a major lestone is going to effect. Peru´s Ministry of Culture, is officially recognizing curanderismo based on the use of San Pedro as cultural patrimory of the nation. Two more symposia, Puemape 2004 and 2005 followed. Building on these precedents, in 2011, the Universidad Privada Antenor Orrego organized an international forum on *curanderismo,* which included *curanderos,* scholars, artists, and public health professionals (Paz, ed. 2012). As this book goes to press, a major milestone is being proposed, Peru´s Ministry of Culture is in legislation declaring north-coastal Curanderismo based on the use of San pedro as "cultural patrimony of the nation"

Hopefully, the foregoing initiatives will lead to additional innovations, including the actual inclusion of traditional practitioners as peers in mainstream therapy and service delivery. For examples of the equitable incorporation of traditional healers into modern health care, one need only look to Peru's neighbors. To the north in Ecuador, where the 2008 Constitution guarantees the rights of indigenous populations to their traditional forms of medicine, a native *yachak* works in a hospital in Riobamba alongside modern health care personnel (Caselli 2012). To the south in Bolivia, indigenous *yatiris* and *kallawayas* are organized in guilds (G. Fernández 2012: 330-334). In Chile, the Mapuche not only administer their regional hospital, but have also organized natural pharmacies which dispense validated medicinal plant remedies on an international basis (*Nuestra Farmacia* 2004). Is it too much to hope that the same can happen in Peru?

Acknowledgements

Fieldwork for this project started initially in 2001 as a collaboration between the MIRT/MHIRT program of San Diego State University (SDSU, Dr. Robert Pozos, PD; Since 2014 PD is Dr. Dena Plemmons), the San Diego Museum of Man (Dr. Douglas Sharon) and the University of Bayreuth, Germany (Dr. Rainer W. Bussmann). From 2003 it was continued under the auspices of SDSU (Pozos), the Phoebe Hearst Museum of Anthropology at UC Berkeley (Sharon) and the University of Hawaii at Manoa (Bussmann). As of 2007 the botanical work moved to the William L. Brown Center at the Missouri Botanical Garden (Bussmann). Since 2010 we have been fortunate to have the collaboration of Dr. Thomas Love (Linfield College) and since 2011, Dr. Gail Willsky (University at Buffalo-SUNY).

Fieldwork for this project was supported through the assistance of MIRT/MHIRT students Marisa Alvarez, Maria Brodine, Christina Dennis, Gabriel Chait, David Dubose, Vanessa Feregrino, Mirna García, Gabriel Guerrero, Erika Hernández, Douglas Highfill, Kasee Houston, Brit Nilsen, Antonio Olea, and James Saleda (San Diego SU); Guadalupe Ochoa (San Fran SU); Grace Kim (U CA SD); Tahamara Ibarra (Truman SU); Lorena Alvarez and Susana Fajardo (Linfield Coll); Ricardo González, Gletys Montoya, and Tamia Souto (U Hawaii-Manoa)); Guy Banner (Utah SU); Cynthia Aguirre, Abelardo Arellano, Yasmin Barrocio, Rosalie Cardenas, Lucy Carrillo, Mayra Castro, Rosa Chan, Rigoberto Del Toro, Taisha Ford, Rubi Galarza, Mirna Garcia, Daisy González, Gladys Guardado, Ana Jones, Cindy Lau, Alexis López, Jennifer Ly, Mariana Martinez, Ryan Martinez, Mariela Nevarez, Jessica Ortiz, María Perez, Tahirah Rasheed, Samuel Regalado, Giselle Rodriguez, Daniel Ruiz, Jaime Sarria, Roberto Silva and Melinda Soriano (U CA Berkeley); Javier Blanco and Shant Tamazian (U Buffalo SUNY).

The following students of medicine provided invaluable service as teaching assistants: Brian Jonat, Kamron Pourmand, Brent Parris, and Keith Willner (SUNY Stony Brook) and Abby Besch, Rachel Ochotney, and Rachel Simpson (U Buffalo – SUNY). Biochemistry volunteers included: Kristen Brooks, Lauren Carnevale, Thomas Cleland, Canaan Coppola, Russell Van Coevering, Lauren VanGelder, and Brian Zylinski (U Buffalo-SUNY) as well as Emily Bacaj (Barnard Coll), Joy Barranis (U Chicago), Emily Frisch (Chapman Coll), Carolyn Gigot (Harvard) and Elizabeth Pon (Brown U); Anthropology/Sociology volunteers (Linfield Coll.) included Sandra Garcia, Sam Gauksheim, Katelyn Henson, Joshua Ness, Kiana Ringette, Ana Sours, and Charlotte Trowbridge - tutored in the field by their department chair, Dr. Thomas Love and Botanist Dr. John Syring.

Three Masters theses were completed on topics related to the project by Erika Hernández (Lat Am Stud, SDSU), Zachary Revene (Anthro, Utah SU), and Ashley Smallwood (LAS, SDSU). Senior theses were written by Jay Chicoine (Anthro, Univ S Dakota) and Guadalupe Ochoa (Anthro, SFSU), and an M.A. analytical essay was submitted by Amanda Zenick (LAS, SDSU).

None of the work would have been possible without the invaluable collaboration of our Peruvian colleagues, *curanderas* Julia Calderón, Isabel Chinguel, and Olinda Pintado; *curandero* Leoncio Carrión; *curandero* Germán Santisteban; and herbalists Manuel Bejarano, Elmer Cruz and Iván Cruz. We especially want to thank them all.

For three field seasons we were fortunate to have the anthropological expertise of our SDSU colleague Alan Kilpatrick. Project collaborators included: Biologist Carolina Téllez and Carlos Vega (INBIA); Microbiologists Marisela Aguilera and Doris Diaz (Clinica Anticona); Biochemists Inés Castro, Gonzalo Malca, and Alan Mayanga (Universidad Nacional de Trujillo), Lab Technician Lai-Har Chi (U Buffalo-SUNY); Botanists Ashley Glenn, Jessica Griffard, Karen Meyer, Alyse Kuhlman, and Andrew Townesmith (William L. Brown Center, Missouri Botanical Garden); volunteers Jay Chicoin, Genevieve Gilbreath, Pablo Lozano, Carlos Reyes, Jason Schoneman, and Selma Somogny; and IT specialists John Effio and Steven Scokzen. A special debt of gratitude goes to our dedicated volunteers Kaye Sharon, Tucker Sharon, and René Summerfield. Since 2011 we have been fortunate to have the professionalism of Dr. Gail Willsky (U Buffalo-SUNY) who has transformed our laboratory methodology and training at UNT with the strong support of Dean of the Faculty of Chemical Engineering Manuel Vera, Ingeniero Químico Alberto Quezada, and Químico-Farmacéutico Mayar Ganoza.

Thanks also go to Eric Rodriguez (Herbarium Truxillense, HUT) and Segundo Leiva, Mario Zapata, and the late Dr. Abundio Sagestegui (Herbario Antenor Orrego, HAO) for the use of their facilities and assistance in plant identification. Dr. Fredy Pérez (UPAO) has been a pioneer phytochemist and tireless tutor of our students in the lab. MD Sarah Cohen (Lakenau PA Health) provided valuable laboratory assessments. From the very beginning we have been able to count on the unstinting good will of our MD colleagues Noé Anticona (Clinica Anticona), Luís Fernández (EsSalud/CAMEC-Trujillo), and the late Hernán Miranda (UNT).

The support of this work under the MIRT/MHIRT grant program (# G0000613, Fund 54112B MHIRT and NIH/NIMHD # 5 T37 MDOO1442-18) of the National Institutes of Health (NIH), the Deutsche Forschungsgemeinschaft (DFG), and the William L. Brown Center the Missouri Botanical Garden are gratefully acknowledged.

Most of all, we want to express our sincere gratitude to the people of Northern Peru for sharing their ethnobotanical knowledge with us.

The Plant

Aphelandra cirsioides

Sambucus peruviana

Tetragonia crystallina

Bomarea angulata

Bomarea dulcis

Alternanthera brasiliana

ACANTHACEAE - *Aphelandra cirsioides* Lindau
Espina de Hoja
Shrub, Andean, 2000-3500m
Use: Bronchitis / Whole plant, dried / Oral / 2 tbsp with 1 liter of boiled water, 3 cups a day, 3-4 days

ADOXACEAE - *Sambucus peruviana* Kunth
Cinta de Novia, Sauco, Saucotillo, Tilo
Tree, Andean, 2000-4000m, cultivated
Use: 1. Swelling, Kidneys, Cough, Concussions, Prostate, Fever, Bronchitis, Yellow Fever / Leaves, Flowers and Stems, fresh or dried / Tea: 5-20g per 1 liter boiled for 1 minute and combineed with Sugarcane Alcohol. Drink 3 times a week, up to 1 liter a day if needed or until fever passes. Take while cold. **2.** Fright/Susto, Fever, Yellow Fever / Leaves, fresh / Topical / Bath: combine with Nogal, Hierba del Susto, Manzanilla Blanca in a flask of Timolina, 2-4 times a month. Not to be used too much because it is very cold. **3.** Inflammation of the kidneys, Gastritis / Leaves, fresh / Topical / Poultice, once a month. **4.** Nerves, Cough, Cold, Fever, Insomnia / Flowers and Leaves, fresh / Oral / Boil 1 liter of water, then add 10g of Sauco, Manzanilla, Hinojo, Coleo, Ajenco, Toronjil, Pimpinela and Claveles. Cover and let it sit for 2-3 minutes. Patient should drink warm solution, 3-4 cups a day for 1 month. **5.** Amulet. To Bind a person in Love Magic / Leaves, fresh / Topical / Tie a picture on the Stems and wrap it. Pray and spray (fogear) perfume with the appropriate names. Repeat as needed. Used for ritual Amarres (Tying up or Binding someone).

AIZOACEAE - *Tetragonia crystallina* L'Héritier
Hierba de la Sonrisa, Señorita, Ulluco de Gentil, Hierba de la Señorita
Herb, Coastal, 0-500m, introduced weed
Use: 1. Happiness, Tranquility, Fragrance, Good Luck, Spiritual Flowering, Protection, Good Health, Good Fortune, Good Business / Stems, Leaves, Flowers, fresh / Seguro / Standard Seguro mixture. **2.** Inflammation of the molars / Stems, Leaves, Flowers, fresh / Topical / Plaster, chop the Leaf and Stems, apply 2 hours, 2 times a day. **3.** Happiness, Tranquility, Fragrance, Good Luck, Spiritual Flowering, Protection, Good Health, Good Fortune, Good Business / Stems, Leaves, Flowers, fresh / Topical / Alternative mixture for Spiritual Flowering.

ALSTROEMERIACEAE - *Bomarea angulata* Benth.
Cachujillo
Herb, Andean, 300-3500m
Use: Infertility in women / Whole plant, dried / Oral / 5g per 1 cup of boiling water, 1 cup daily for 8 months.

ALSTROEMERIACEAE - *Bomarea dulcis* (Hook.) Beauv.
Espuela de Gallo
Herb, Andean, 3000-4500m
Use: Protection, Success, Advising / Whole plant, fresh / Seguro / 3 Stems per flask

AMARANTHACEAE - *Alternanthera brasiliana* (L.) Kuntze
Hierba del Oso, Veronica (Hembra), Moradilla de Cerro
Herb or shrub, Amazonian, 0-500m, weed
Use: 1. Disjointed Bones, Bronchitis, Asthma, Bruises, Fractures, Bumps / Whole plant, fresh or dried / Oral / 5-10g in 1 liter boiling water. Mix with Muyaca, Huamanripa, Brochamelia. 4 cups a day, 1-2 weeks. **2.** Susto of heights/Susto de la cumbre / Whole plant, fresh or dried / Topical / Limpia or bath. 5g per 3 liters of boiled water. Also use with Hierba del Susto, Zanahoria de Gentil, Poleo de Gentil. 1-2 times a month.

Alternanthera halimifolia

Alternanthera porrigens

Alternanthera villosa

Amaranthus caudatus

Amaranthus hybridus

Chenopodium ambrosioides

AMARANTHACEAE - *Alternanthera halimifolia* (Lam.) Standley & Pittier
Paja Morada (Colores), Lenguetilla, Sanguinario, Lengua de Pájaro, Sanguinaria, Moradilla, Hierba Morada
Herb, Amazonian and Andean, 0-2500m, weed
Use: 1. Inflammation / Whole plant, fresh / Oral / Boil 1 liter of water, add 10g Paja Morada. Combine with Malva, Pie de Perro, Cola de Caballo, Chacur and Sombrerito. Drink the solution and use it as a wash, 3 times a day as needed. **2.** Inflammation / Whole plant, fresh / Topical / Boil 1 liter of water, add 10g Paja Morada. Combine with Malva, Pie de Perro, Cola de Caballo, Chacur, and Sombrerito. Drink the solution and use it as a wash, 3 times a day as needed. **3.** Nerves, Heart, Anxiety, Heart disease, Relaxation / Whole plant, fresh / Oral / Boil 1 liter of water, then add 10g total of Toronjil, Manzanilla, Romero, Hinojo, Chanca de Comida, Cascorade, and Membrillo. Let mixture sit for 2-3 minutes. Drink lukewarm 3-4 times a day during meals, or as needed. Patient should drink cold solution. It is important to drink it at 6AM and at 6PM.

AMARANTHACEAE - *Alternanthera porrigens* (Jacquin) Kuntze
Sanguinaria, Moradilla, Lancetilla
Shrub, Amazonian, Andean and Coastal, 0-2500m, weed
Use: 1. Blood circulation, Warts, Cold Blood, Allergies / Whole plant, fresh or dried / Oral / 5g in 1 liter of water combined with Conchalagua, Moradilla, Colores, Lancetilla, Culantrillo, Hierba del Toro and Zarza Parilla. Drink 3 times a day up to one year. **2.** Cleansing womb after childbirth, Fragrance, Luck in love and work, Bad Air/Mal Aire, Love, Business Relations, Protection, Good Fortune, Good Health / Whole plant, fresh or dried / Topical / Alternative mixture for Spiritual Flowering. Take 3 baths a month. **3.** Good Business, Protection, Good Fortune, Good Health / Whole plant, fresh or dried / Seguro / Standard Seguro mixture.

AMARANTHACEAE - *Alternanthera villosa* Kunth
Hierba del Oso
Andean, 1000-2000m, weed
Use: 1. Bad Air/Mal Aire, Sorcery (protection from) / Flowers, Leaves and Stems, fresh or dried / Oral / Boil 20g of Hierba del Oso in 1/2 cup of water for 5 minutes. Boil 20g of Hierba del Oso in 1/2 cup of water for 5 minutes. Drink cold, 1/8 cup once only. **2.** Protection from evil / Flowers, Leaves and Stems, fresh or dried / Topical / Bath mixture for Protection from evil.

AMARANTHACEAE - *Amaranthus caudatus* L.
Quihuicha, Kiwicha
Herb, Coastal, Andean, 0-3500m, cultivated
Use: Nutrition supplement / Seeds, dried / Oral / 150g of the grain and 1 liter of water. Boil for 10 minutes or until grain is soft. Add Cinnamon, Apples, and Membrillo. 1 cup 1-2 times a day as necessary.

AMARANTHACEAE - *Amaranthus hybridus* L.
Yuyo
Herb, Amazonian, Andean, 0-3500m, weed
Use: Inflammation (general) / Leaves and Stems, fresh / Oral / 100g of Yuyo and 1/2 cup of water boiled for 5 minutes. Drink cold, 1/4 cup 2 times a day for 3 days.

AMARANTHACEAE - *Chenopodium ambrosioides* L.
Paico
Herb, Amazonian, Andean, Coastal, 0-4000m, weed
Use: 1. Parasites (worms) / Leaves and Stems, fresh / Oral / Extract the juice of the Leaves. The oil of the Seed and Fruit has an ingredient that kills parasites. Use once a month. **2.** Cough / Leaves and Stems, fresh / Oral / Add 10g of plant material to 1/2 liter of water. Drink hot, 1 cup 2-3 times a day for 1 week.

Chenopodium quinoa

Chenopodium quinoa

Iresine diffusa

Iresine herbstii

Allium odorum

Allium sativum

AMARANTHACEAE - *Chenopodium quinoa* Willd. (wild form)
Quinua Amarga
Herb, Andean, 2000-4000m, weed
Use: Intestines (cleansing), Stomach (cleansing) / Seeds, fresh / Topical / 5g dry or 5g per 1 liter of water mixed with Chocón, once a month as enema.

AMARANTHACEAE - *Chenopodium quinoa* Willd.
Quinua
Herb, Andean, Coastal, 0-4000m, cultivated
Use: Nutritional supplement / Seeds, fresh or dried / Oral / To 1 liter of water add 150g of Quinua and boil for 10 minutes or until the grain is soft. Add a piece of cinnamon and a piece of apple. Take 1 cup 1-2 times a day as needed.

AMARANTHACEAE - *Iresine diffusa* Humb. & Bonpl. ex Willd.
Paja Blanca, Sanguinaria
Herb or shrub, Amazonian, Andean, 0-3500m, weed
Use: Liver, Kidneys, Inflammation of the Ovaries, Blood, Menstruation symptoms in adolescents / Whole plant, fresh / Oral / Boil 10g in 1 liter of water, and mix with Ambarina, Lancetilla, Hierba de la Rabia and Palo de Sangre. Drink 3 times a day or as needed, 1 liter daily, for 1 year.

AMARANTHACEAE - *Iresine herbstii* Lindley
Colores, Timoras, Zangurache
Amazonian, 0-500m, cultivated
Use: 1. Liver, Kidneys, Cancer of the blood, Blood circulation, Intoxication of the blood, Heart, Nervous system, Blood, Inflammation of the Stomach, Inflammation (general) / Leaves, fresh / Topical / Fresh Leaves only, may use with Sugarcane Alcohol, Vinegar, and Contrahierba. As poultice, 3 times a week. **2.** Liver, Kidneys, Cancer of the blood, Blood circulation, Intoxication of the blood, Heart, Nervous system, Blood, Inflammation of the Stomach, Inflammation (general) / Leaves, fresh / Oral / Boil 5g in 1 liter of water with Lancetilla, Contrahierba, Cachorillo and eat fresh. Drink once a day for 1 week to 1 month, always before breakfast.

AMARYLLIDACEAE - *Allium odorum* L.
Cebolla China, Cebolla (Onion)
Herb, Andean, Coastal, 0-3500m, introduced and cultivated
Use: 1. Bronchitis, Asthma / Whole plant, fresh / Oral / Dice 15 onions in a bowl. Add a glass of water and 1/4kg of white sugar. Add a piece of ginger (can also add hen fat). Boil and stir until thick. Drink syrup, 5g every 6 hours for 1 week. Juice can also be drunk naturally. **2.** Bruises, Bad Air/Mal Aire, Blood clots / Whole plant, fresh / Topical / Crush 1 1/2kg of Cebolla and strain in a piece of cloth to obtain all the extract. Discard the juice and use the rest. Place on top of the affected area and cover with a piece of cloth, every other day in the morning for 3 days.

AMARYLLIDACEAE - *Allium sativum* L.
Ajo (Garlic)
Herb, Andean, Coastal, 0-3500m, introduced and cultivated
Use: 1. Cough, Bronchitis, Cold / Clove, fresh / Oral / Place 3 garlic cloves, 1 Chinese onion, Matico, Escorcionera, Eucalyptus, Vira Vira, white sugar, and 1/2 liter of water or cow's milk in a pot and boil for 3 minutes. Drink warm, 2 tbsp twice a day for 1 week. Can also be eaten raw. **2.** Bruises, Arthritis, Rheumatism, Bad Air/Mal Aire / Clove, fresh / Topical / Crush 250g of garlic. Add it to 10g Eucalyptus, 90g of Alcohol, Pacra, Chuchuhuasi and Ginger. Let these ingredients soak in 1 liter of alcohol for 1 week. Massage and rub the mixture on affected areas. Rub 1-2 times a day as needed. **3.** Bad Air/Mal Aire, Removing bad spirits from the house / Peel, fresh / Incense / Burn 1kg of the peel in top of a char-

Eustephia coccinea

Anacardium occidentale

Loxopterygium huasango

Mangifera indica

Mauria heterophylla

Schinus molle

AMARYLLIDACEAE - *Eustephia coccinea* Cav.
Tumapara, Pomanpara, Puma Para, Para Para
Herb, Andean, 2000-400m, cultivated
Use: 1. Arthritis, Rheumatism / Bark, fresh or dried / Oral / Macerate in a bottle of wine, 3 small glasses a day. **2.** Inflammation of the Stomach, Undo Witchcraft / Bark, dried / Oral / Boil 200g of Pomanpara in 1 liter of water for 10 minutes. Drink cold, 1 cup every 3 days for 1 month. **3.** Wounds / Bark, dried / Topical / Crush and pulverize with a rock, then drain to create a powder. Place powder on top of the wound, once a day until the wound is healed. Boil for 20 minutes, 20g of herb in 1 liter of water mixed with Matico, Malva and Talla. Wash wound once a day for 8 days. **4.** Inflammation, Hemorrhages, Inflammation of Uterus, Ulcers, Cysts, Cancerous wounds / Bark, fresh or dried / Oral / Boil 3-5 minutes, 5-10g in 1 liter of water mixed with Flor Blanca, Purenrosa, Malva Olorosa. Drink 3 times a day for 8 days.

ANACARDIACEAE - *Anacardium occidentale* L.
Marañon, Cayu (Cashew)
Tree, Amazonian, Andean, 0-1000m, cultivated
Use: Scars, Moles, Cysts (ingrown), Skin stains / Seeds, fresh / Topical / Crush seeds and collect "blood." Topical application on affected area.

ANACARDIACEAE - *Loxopterygium huasango* Spruce ex Engl.
Hualtaco
Tree, Amazonian, Andean, 0-1000m
Use: Bone or muscular pain due to an accident, pain from bone fractures / Wood, dried / Oral / Boil 20g of Hualtaco with Diego Lope, Suelda con Suelda, and 1 liter of water for 30 minutes. Patient should drink cold solution. Do not exceed 3 dosages of the treatment because it is very strong. 1/2 cup a day every other day, or as needed.

ANACARDIACEAE - *Mangifera indica* L.
Mango
Tree, introduced and cultivated
Use: Bronchitis, Colds, Inflammation (chest) / Leaves, dried / Oral / Boil 5 Mango Leaves with 10 Moy Leaves, 10 Eucalyptus Leaves, 5 Stems of Pájaro Bobo, and 1 Lemon Leaf (all dried Leaves) in 1 liter of water for 30 minutes. Drink cold, 2 tbsp. twice a day for 3 days.

ANACARDIACEAE - *Mauria heterophylla* Kunth.
Shimir, Tres Hojas, Trinidad, Chacur, Ahimir, Feregreco
Tree, Andean, 500-4000m
Use: 1. Daño/Sorcery, Fright/Susto, Skin irritation from Daño/Sorcery / Leaves, fresh / Topical / Boil 50g with Lailambo, Nogal, Ajenco, Timolina. Limpia: once a week. **2.** Inflammation, Liver, Kidneys, Wounds, Inflammation of Uterus, Cleansing (external), Cleansing (internal), Ulcers (internal), Ulcers (external), Inflammation of the Ovaries, Cysts, Fibroids / Leaves, fresh / Oral / 10g per cup, combined with Cola de Caballo, Verbena, and Amor Seco. Drink the solution and use as a wash. Take 4 cups a day for 1 month. **3.** Vaginal cleansing / Leaves, fresh / Topical / 1/2 liter per 1 Stem with Leaves. Situate legs in a "V" position and drop solution into vagina for 10 minutes. Go to restroom and contract muscles till fluid has drained. Repeat if necessary, twice a month.

ANACARDIACEAE - *Schinus molle* L.
Molle, Moy
Tree, Amazonian, Andean, Coastal, 0-3500m
Use: 1. Arthritis, Rheumatism, Bone pain, Bronchitis, Cough, Cold, Chills, Inflammation (external) / Flowers, Leaves and Stems, fresh / Topical / Macerate material in alcohol and spray on patient at nighttime, once daily for 5 days as poultice or rub the patient's body with plant material while bathing in the

Annona muricata

Ammi visnaga

Apium graveolens

Arracacia xanthorrhiza

Coriandrum sativum

Daucus montanus

mixture. Advise the patient to rest and to avoid going outdoors. **2.** Arthritis, Rheumatism, Bone pain, Bronchitis, Cough, Cold, Chills, Inflammation (external) / Flowers, Leaves and Stems, fresh / Oral / 20g crushed and mixed with alcohol. Boil 20-30 hours mixed with Eucalyptus, Ruda, Chamana, and Tilo. Take 1 cup, 4 times a day for 2 months or as needed. **3.** Cancer, Tuberculosis / Bark and Latex, fresh / Oral / Add 20g of Bark resin (approximately 5cm) to 1 liter of water. Boil the combination for 3 minutes. Take 1 cup, 4 times a day for 2 months or as needed. **4.** Vaginal infection / Bark and Latex, fresh / Topical / Add 20g of Bark resin (approximately 5cm) to 1 liter of water. Boil the combination for 3 minutes. Use 1 cup, 4 times a day for 2 months as vaginal douche.

ANNONACEAE - *Annona muricata* L.
Guanábana, Graviola
Tree, Amazonian, 0-500m, cultivated
Use: Gastritis, Inflammation, Kidneys, Cancer / Leaves, fresh / Oral / Boil 1/2 liter of water with 10 Leaves of Guanábana, 10g of Amor Seco, Pineapple Peels and Achiote for 3-4 minutes. Drink cold, 3-4 cups a day for 1 month.

APIACEAE - *Ammi visnaga* (L.) Lam.
Visnaga
Herb, Coastal, Andean, 0-3000m, introduced
Use: Bad Air/Mal Aire, Headache / Flowers and Leaves, fresh / Topical / 20g crushed Leaves as a Poultice, or 20g per 5 liters of water for 20 minutes as a Bath, 3 times a week.

APIACEAE - *Apium graveolens* L.
Apio Cimarrón, Apio
Herb, Andean, Coastal, 0-3000m, introduced
Use: 1. Colic, Bronchitis, Heart, Nerves, Insomnia, Anxiety, Gases, Gastritis, Colic of the Stomach / Whole plant, fresh / Oral / Boil 1 liter of water, add 10g of Apio Cimarrón. Combine with Manzanilla, Mejorana and Culantrillo. Drink 4 cups a day for 1 week. **2.** Fright in children / Susto en niños, Gastritis / Whole plant, fresh / Topical / Boil with Perejil. Mix with Agua del Susto, 3 baths per month.

APIACEAE - *Arracacia xanthorrhiza* Bancroft
Racacha, Racacha Cimarrona
Herb, Andean, 3000-4000m, weed
Use: Fright/Susto / Leaves and Stems, fresh / Topical / Boil 20g of plant material with Flor de Chocho, Eucalyptus, Chueguis and 2 liters of water. Bathe the patient in the warm mixture while rubbing with the Leaves. Bathe 3 times a week for 1 month using 1 cup of material.

APIACEAE - *Coriandrum sativum* L.
Culantro
Herb, Amazonian, Andean, Coastal, 0-3500m, weed, introduced
Use: Bad Air/Mal Aire that takes your sight / Leaves, fresh / Topical / Place fresh Leaves over eyes. Apply only once. Leave for 1 hour.

APIACEAE - *Daucus montanus* Humb. & Bonpl. ex Spreng.
Zanahoria de Zorro, Zanahoria de Gentil, Zanahoria
Herb, Andean, Coastal, 0-4500m, weed
Use: 1. Bad Air/Mal Aire, Sorcery, Twists caused by Sorcery, Twists, Contusions caused by Sorcery, Fright from Ruins/ Susto de Huaca, Fear of Heights / Susto de las Cumbres / Leaves and Stems, fresh / Topical / Leaves with Vinegar and 7 Espíritus, 1 bundle in 3 liters of boiled water with Ishpinguillo, Conchalay Blanco, Manzanilla de Cerro Lailambo and Timolina. As Bath, Limpia or Poultice, 1-2 times a week. **2.** Eyesight, Sunspots / Root, fresh / Topical / Grate root, 3 drops in each eye or on affected areas of skin, 2 times a day for 2-3 days. **3.** Inflammation (general) / Root, fresh / Oral / 50g of the tuber and 1/4 cup of water, blend and strain. 1 glass once a day for 15 days. Drink cold during breakfast while fasting.

Foeniculum vulgare

Niphogeton dissecta

Petroselinum crispum

Pimpinella anisum

Mandevilla antennacea

Mandevilla trianae

APIACEAE - *Foeniculum vulgare* P. Miller
Hinojo, Anís Criollo
Herb, Andean, Coastal, 0-2000m, weed, introduced
Use: 1. Colic, Gases, After birth, Stomach Pain, Stomachache, Nerves, Diarrhea / Whole plant, fresh / Oral / Boil 5g in 1 liter of boiling water. Combine with Manzanilla, Poleo, Toronjil, Pimpinela, Clavel and Borraja. Drink 3 times a day for 1 month. **2.** Colic, Gases, After birth, Stomach Pain, Stomachache, Nerves, Diarrhea / Seeds, fresh / Oral / Add 1 tsp. of plant material to 1 cup of water. Boil mixture for 2 minutes. Drink warm. Honey or sugar can be added, if desired, twice a day for 2 days.

APIACEAE - *Niphogeton dissecta* (Benth.) J.F. Macbr.
Hórnamo Toro
Herb, Andean, 4000-4500m
Use: 1. Wounds (cancerous), Wounds from Sorcery / Leaves and Stems, dried / Topical / Boil 20g per 5 liters of water for 20 minutes and mix with other Hórnamos. Bath, 3 times a week. **2.** Purgative / Whole plant, fresh / Oral / Boil 5g per 1/2 liter of water and drink 1 cup a day for 1 month.

APIACEAE - *Petroselinum crispum* (Miller) A.W. Hill
Perejíl
Herb, Andean, 3000-4500m, weed, introduced
Use: 1. Heart, Nervous system, High blood pressure, Infections, Nosebleeds, Food condiment, Forgetting love or trauma, Regulation of menstrual cycle / Whole plant, fresh / Oral / 3-5g of herb in 1 liter of water combined with Toronjil, Pimpinela, Mejorana, and Siempre Viva. Take 1 glass 2 times a day for 3 days before the menstrual period and three days after. Can also be eaten as salad. **2.** Daño/Sorcery, Fright/Susto / Whole plant, fresh / Topical / Bundle fresh Leaves and Stems with Apio and burn. **3.** Infections, Nosebleeds, Food condiment, Forgetting love or trauma / Whole plant, fresh / Topical / Crush herb and boil with meat and salt. Apply as poultice, twice a month as Limpia or Bath.

APIACEAE - *Pimpinella anisum* L.
Anis Criollo, Anís
Herb, Andean, Coastal, 0-3000m, weed, introduced
Use: Gases, Stomach Pain, Colic / Seeds, dried / Oral / Tea: 5-20g in 1 liter of water with Menta and Manzanilla, 2-3 cups a day for 3 days or as needed.

APOCYNACEAE - *Mandevilla antennacea* (A.DC.) Schum.
Bejuco Colambo Negro
Vine, Amazonian, Andean, 0-1500m
Use: Protecting the house and field / Whole plant, fresh / Amulet / Plant close to the house.

APOCYNACEAE - *Mandevilla trianae* Woodson
Bejuco, Bejuco Negro (Grande), Bejuco Negro (Chico)
Vine, Amazonian, Andean, 0-1000m
Use: Daño/Sorcery de Brebaje, Fright/Susto, Sorcery / Leaves, fresh or dried / Topical / 1 handful per 3 liters of boiled water. Can combine with Zanahoria de Gentil, Chilca, Añasquero Chico, Ishpinguillo, Conchalay, Hierba del Susto (when used for Susto) and 7 Espiritus. Bathe once a week and have a Limpia once a month.

Nerium oleander

Thevetia peruviana

Vallesia glabra

Ilex guayusa

Hydrocotyle bonariensis

Hydrocotyle globiflora

APOCYNACEAE - *Nerium oleander* L.
Laurel, Laurel Rosa
Shrub, Amazonian, Andean, Coastal, 0-1500m, introduced, cultivated
Use: Cleaning wounds, Itching, Sarna, Rashes, Skin Lesions, Herpes / Flowers, Leaves and Stems, fresh / Topical / Boil 30g Laurel in 5 liters of water for 5 minutes. Use water to shower. Rub Leaves on the skin. Don't touch the eyes or mouth while washing because the Leaves are poisonous. Wash every other day for 7 days or 3 times: Tuesday, Friday, Tuesday.

APOCYNACEAE - *Thevetia peruviana* (Pers.) Schum.
Mailchin, Maichil, Camalonga, Cabalonga
Shrub, Amazonian, Andean, 0-2500m, weed
Use: 1. Bones, Rheumatism, Bad Air/Mal Aire, Arthritis, Bad Luck Spell (Sorcery) / Stems and Leaves, fresh / Topical / Bath: add 10g of plant Leaves to 2 liters of water and boil the mixture for 3 minutes, or boil 20 minutes in 5 liters of water. Apply the mixture lukewarm. Rub the patient with Flowers and water. After bath, advise patient to dress in warm clothing, 2 baths a week (Tuesday and Friday) or 3-4 times a month. Do not ingest! **2.** Bad Air/Mal Aire, Epilepsy, Nerves, Heart Attack / Seeds, dried / Oral / Ground and boiled, best macerated in wine that has low alcohol content. Must use holy wine from the church. Mixed with seeds from seven other plants: Ashango, Pucho, Amala, Quina Quina, Nuez Mozcada, and Ishpingo, once a month or as needed. **3.** Menopause, Cancer, Bad Air/Mal Aire, Sorcery/ Daño / Seeds, dried / Oral / Place 1 seed of Camalonga in 1 bottle of wine and let sit for 8 days, drink 1 small wine of glass once a day for 20 days or as needed.

APOCYNACEAE - *Vallesia glabra* (Cav.) Link.
Cuncuno, Cun Cun
Tree, Coastal, 0-1000m
Use: 1. Snake bites / Leaves, fresh / Oral / Boil 15 Leaves of Cuncuno with 10 seeds of Faique and 1/8 liter of cooking oil in 1 liter of water for 20 minutes. Drink cold, 1/2 cup twice a day (6 AM and 6 PM) for 2 days. Do not eat sea food or spices (no chili peppers) and stay away from the sunlight during treatment. **2.** Diabetes / Leaves, fresh / Oral / Blend 15 Leaves until you have an extract. Patient should drink cold solution. Drink only at 6 AM. Do not eat anything sweet while on treatment, 1 small glass every morning for 30 days.

AQUIFOLIACEAE - *Ilex guayusa* Loes
Guayusa, Agracejo, Citrodora
Shrub, Andean, 1000-1500m
Use: Diabetes, Intoxication of the blood / Leaves, dried / Oral / Tea, 5-10g in 1 liter of water, 1 cup 3 times a day as needed. Drink warm.

ARALIACEAE - *Hydrocotyle bonariensis* Commerson ex Lam.
Tutapure de Estrella
Herb, Andean, Coastal, 0-4500m, weed
Use: Sorcery / Leaves and Stems, dried / Topical / 1 handful boiled in 3 liters of water. Can combine with Bejuco Amarillo and Palo Blanco, 1 bath per month.

ARALIACEAE - *Hydrocotyle globiflora* R. & P.
Sombrerito
Herb, Andean, 1000-2500m, weed
Use: Liver, Inflammation of the Kidneys / Whole plant, fresh / Oral / 4 small Leaves and 2 Flowers in 1 liter of water boiled for 3 minutes. Drink 3 times a day for 1 month.

Oreopanax eriocephalus

Araucaria heterophylla

Bactris gasipaes

Cocos nucifera

Aristolochia ruiziana

Sarcostemma clausum

ARALIACEAE - *Oreopanax eriocephalus* Harms
Maque Maque, Mano de León
Shrub, Andean, 2500-3500m
Use: 1. Heart, Nerves, Inflammation, Fractures, Nervous system, Rheumatism, Protection from evil / Leaves and Flowers, fresh / Oral / Boil 3-10g in 1 liter of water. Drink 3-4 times a day for 1 month. **2.** Heart, Nerves, Inflammation, Fractures, Nervous system, Rheumatism, Protection from evil / Leaves and Flowers, fresh / Topical / Boil 3-10g in 1 liter of water. Take as bath. **3.** Fright/Susto / Leaves and Flowers, fresh / Topical / 5g in 3 liters of boiling water mixed with Laurel, Rumilanche, Poleo de Gentil, 7 Espiritus, Manzanilla Blanca, Romero Castillo, and Nogal. Bathe 3 times a month.

ARAUCARIACEAE - *Araucaria heterophylla* (Salisb.) Franco
Pino
Tree, Andean, Coastal, introduced and cultivated
Use: Toothache, Extracting teeth / Resin, fresh or dried / Topical / Warm up a small piece of resin. Place resin on top of the tooth, 2 times a day as needed or until tooth is healed.

ARECACEAE - *Bactris gasipaes* Kunth
Chonta
Tree, Amazonian, 0-500m, cultivated
Use: Protection / Wood / Amulet / Pass over body while praying.

ARECACEAE - *Cocos nucifera* L.
Coco
Tree, Amazonian, Coastal, 0-500m, introduced and cultivated
Use: Diarrhea, Parasites, Inflammation of the liver, Inflammation (general) / Fruit Peel, dried / Oral / Grind 10 seeds of Coconut and boil in 1/4 liter of water for 20 minutes combined with 1-2 Leaves of Hierba Luisa, Culen, Hinojo, and Poleo. Drink 3 times a day for 2-3 days. Drink the coconut milk for inflammations.

ARISTOLOCHIACEAE - *Aristolochia ruiziana* (Klotsch) Duch.
Bejuco de Contra-Aire
Liana, Amazonian, 0-500m
Use: Untangling a person who is confused or having trouble progressing in life / Stems, dried / Oral / Boil 20g of Bejuco in 1 cup of water for 10 minutes. Patient should drink cold solution, once only.

ASCLEPIADACEAE - *Sarcostemma clausum* (Jacquin) Schultes
Marrajudio
Vine, Amazonian, Andean, 0-2000m, weed
Use: 1. Fright/Susto, Rashes, Pimples, Cold sores, Skin marks / Leaves, fresh / Topical / To 2 liters of water add 10g of Flor de Retama, Quinual, Flor de Chuco, and 20g of Eucalyptus. Boil for 3 minutes. Let it cool. Once cooled take a bath, 2-3 times a week or as needed. Alternatively break a Stem and collect the Resin. Apply to affected area, twice a day (AM and PM) as needed. **2.** Promoting lactation in women after birth / Leaves, fresh / Oral / Boil 5 Leaves and 1 Stem of a plant in 1/2 liter of water for 10 minutes. Drink cold, 1 small cup 3 times a week, AM only.

Dracaena fragrans

Acanthoxanthium spinosum

Achillea millefolium

Achyrocline alata

Acmella ciliata

Ambrosia arborescens

ASPARAGACEAE - *Dracaena fragrans* Ker Gawl.
Flor de Dracena
Tree, Andean, Coastal, 0-2000m, introduced
Use: Cough, Bronchitis, Asthma / Leaves and Stems, fresh or dried / Oral / Boil 10g in 1 liter of water, 3 cups.

ASTERACEAE - *Acanthoxanthium spinosum* (L.) Fourreau
Juan Alonso, Espina de Perro, Corona de Cristo
Herb, Andean, 2000-3500m, weed
Use: Detoxification of alcohol and drugs, Inflammation, Bronchitis, Hemorrhages / Whole plant, fresh or dried / Oral / Boil 10g of plant material with 1 liter of water for 3-5 minutes. Also add Alcaparrilla and Guava seeds. Drink warm, 1-2 liters a day, for 2-3 months.

ASTERACEAE - *Achillea millefolium* L.
Milenrama, Chonchon
Herb, Andean, 2500-3500m, introduced and cultivated
Use: 1. Gastritis, Diabetes, Blood, Cholesterol / Flowers and Leaves, fresh / Oral / Boil 3-5g in 1 liter of water and drink 3 times a day for 1 week. **2.** Skin infection, Dispel Spells / Flowers and Leaves, fresh / Topical / 1 bundle in 5 liters of boiling water, 3 baths a month in the evening.

ASTERACEAE - *Achyrocline alata* (Kunth) DC.
Ishpinguillo, Ishpingo, Flor de Ishpingo
Herb, Andean, 2000-4500m, weed
Use: 1. Fright/Susto, Daño/Sorcery in children, Arthritis, Bones / Stems and Leaves, dried / Topical / Bath and Limpia: 5g boiled in 3 liters of water mixed with Añasquero Grande, Ajenco and Tres Hojas, 2 times a month. Alternatively, 1/2 g boiled 10 minutes for a steam bath. **2.** Arthritis, Bones / Stems and Leaves, dried / Topical / Poultice: Use 1 bundle of fresh Leaves with 7 Espiritus and Vinegar for 6 hours, twice a month or as long as illness requires.

ASTERACEAE - *Acmella ciliata* (Kunth) Cass.
Ufla
Herb, Amazonian, Andean, Coastal, 0-3000m, weed
Use: Internal bleeding, Renal bleeding, Cold with high mucus / Root, dried / Oral / Boil 100g of Ufla root and 100g of Menta in 1 liter of water for 10 minutes. Patient should drink lukewarm solution 2 times a day for 3 days.

ASTERACEAE - *Ambrosia arborescens* Mill.
Ambrosia
Herb, Andean, Coastal, 500-4500m, weed
Use: 1. Spiritual Flowering / Whole plant, fresh / Topical / Alternative mixture for Spiritual Flowering. One time only. **2.** Good Business, Protection, Good Fortune, Good Health / Whole plant, fresh / Seguro / Standard Seguro mixture.

Ambrosia peruviana

Arctium lappa

Aristeguietia gayana

Arnica montana

Artemisia absinthium

Ayapana amygdalina

ASTERACEAE - *Ambrosia peruviana* Willd.
Altamisa, Marco, Artamisa, Manzanilla del Muerto, Marcos, Alta Misa, Ajenco, Altamis, Llatama Negra Malera, Llatama Roja Malera
Herb, Amazonian, Andean, Coastal, 0-1500m, weed
Use: 1. Heart, Nerves, Epilepsy, Liver, Bronchitis, Colds, Bad Air/Mal Aire, Burns / Leaves and Stems, fresh / Oral / Boil in 1 liter of water for 2 minutes, then mix water with a total of 10g of Manzanilla, Madre Selva, Hinojo, Borraja, Madre Selva, Toronjil, Manzanilla, Hinojo and Chancas de Comida for Nerve Disorders. Use Boldo, Malva and Linaza for Liver ailments. Use Matico, Borraja, Eucalyptus, Vira Vira, and Brochamelia for Bronchitis. Cover and let sit for 2-3 minutes. Drink lukewarm, 3-4 cups a day for a month. Colds: Boil 1/2 liter of water with 50g of Altamiz and 10g of Sauce, Chicoria, and Pájaro Bobo for 10 minutes. 2 tbsp every 8 hours for 8 days. **2.** Fungus, Fright/Susto / Leaves and Stems, fresh / Topical / Boil 200g in 3 liters of water, 7 Espiritus and Agua de Susto. Use as Poultice or Bath, 6 hrs per Bath, 2 times a month for 1 month; for Susto 3 times a week: Tuesday, Friday, and the following Tuesday. **3.** After birth to reduce inflammation and prevent spasms in womb / Leaves and Stems, fresh / Topical / Poultice: Crush 200g of Leaf and add 5 drops of Trementina (Turpentine). Place Poultice on affected area (woman's abdomen) and cover with a piece of cloth. Leave for 2 hours, 2 times only 2 days apart.

ASTERACEAE - *Arctium lappa* L.
Lampazo
Herb, Andean, Coastal, 0-3000m, introduced
Use: Urinary problems, Skin, Liver, Gallbladder, Intestine, Tumors / Seeds, dried / Oral / Boil for 5 minutes 1/2 liter of water with 10g of Cadillo, Amor Seco, and Triñozo. Drink lukewarm, 1-2 cups 3 times a day for 20 days as needed.

ASTERACEAE - *Aristeguietia gayana* (Wedd.) R.M. King & H. Rob.
Asma Chilca, Asma Chica
Shrub, Andean, Coastal, 2000-3000m
Use: 1. Cough, Bronchitis, Asthma / Leaves, fresh / Topcial / Poultice: 200g with Balsamo de Buddha, 2 times a month. **2.** Cough, Bronchitis, Asthma / Leaves, fresh / Oral / Boil 5g in 1 liter of water mixed with Tilo, Huamanripa, Borraja, and Nogal, 4 cups a day for 10 days.

ASTERACEAE - *Arnica montana* L.
Arnica
Herb, only available as extract, introduced
Use: To wake a person who has fainted, to keep evil spirits away from the house / Stem and Leaves / Amulet / Place close to patient for inhaling. Splash all corners of the house and the center and make a sign of the cross at the front door.

ASTERACEAE - *Artemisia absinthium* L.
Ajenco
Herb, Andean, 2500-3500m, introduced and cultivated
Use: 1. Fright/Susto in children, Sorcery / Leaves, fresh / Topical / Natural, as Limpia, 1 bundle with Vinegar, Añasquero Grande, Añasquero Chico, Flor del Muerto, Hierba del Susto, 7 Espiritus and Agua del Susto, twice a week (Tuesday and Friday). **2.** Fright/Susto in children, Sorcery / Leaves, fresh / Topical / Bath: with Añasquero Grande, Ruda Hembra, 7 Espiritus, and Agua del Susto (if you have Susto), twice a week. **3.** Sorcery / Leaves, fresh / Topical / Bath: with Añasquero Grande, Ruda Hembra, 7 Espiritus, and Agua del Susto (if you have Susto), twice a week. **4.** Menstrual Colics, Menstruation, Regulating the menstrual cycle / Whole plant, preferably Leaves and Stems, fresh / Oral / Tea: 6-10 Leaves per 1 cup of boiling water. 1 cup daily for 3 days or add 100g of the plant to 1 cup of water. Boil the mixture for 5 minutes, take 1/4 cup, once a day, for 3 days.

ASTERACEAE - *Ayapana amygdalina* (Lam.) R.M. King & H. Rob.
Chilco Hembra
Shrub, Amazonian, Andean, Coastal, 0-2000m
Use: Rashes / Leaves and Stems, fresh / Topical / Boil 5 liters of water containing 20g of plant material

Baccharis caespitosa

Baccharis chilco

Baccharis genistelloides

Baccharis latifolia

Baccharis pedunculata

Baccharis salicifolia

ASTERACEAE - *Baccharis caespitosa* (Ruiz & Pav.) Pers
Paja Amargoza
Shrub, Andean, 3000-4000m
Use: Swellings / Flowers and Leaves, fresh / Topical / Grind 500g of material and apply as Poultice, once a day for 8 days on affected area.

ASTERACEAE - *Baccharis chilco* Kunth
Pata de Gallina
Herb, Andean, 1000-3000m
Use: Protection for job and home, Protection (general) / Whole plant, fresh or dried / Seguro / Mix in a bottle 10g of Valeriana Estrella, Señorita, Carpintero, Chupa Flor, Hierba de la Coqueta, Oro, Dolar. Add Agua Florida, Ramillete de Novia, Tabú, and Agua Bendita. Keep on one's person or bedside table.

ASTERACEAE - *Baccharis genistelloides* (Lam.) Pers.
Simba Simba, Carceja, Karqueja, Cadillo
Herb, Andean, 500-4500m
Use: Diabetes, Blood, Cholesterol, Kidneys, Internal Inflammation, Liver, Gallbladder, Baldness, Fat Reduction / Whole plant, fresh / Oral / Boil 10g in 1 liter of water for 2 minutes mixed with Canchalagua, Verbena, Amor Seco, Cola de Caballo, Hierba del Toro, Camote, and Mal Rubio. Take at breakfast and dinner, 1 liter a day for 1 week to 1 month.

ASTERACEAE - *Baccharis latifolia* (Ruiz & Pav.) Pers.
Chilca Chica, Chilca Grande
Shrub, Andean, 1000-4500m
Use: 1. Bone pain, Rheumatism, Arthritis / Leaves and Stems, fresh or dried / Topical / Bath: 5g in 3 liters of water mixed with Manzanilla Blanca, Hierba del Susto, Laurel and Agua del Susto, 2-3 baths per 1 month. Limpia: once a week. **2.** Bone pain, Rheumatism, Arthritis / Leaves, fresh / Topical / Poultice: 200g of fresh Leaves mixed with 7 Espiritus. Apply 1 bundle of fresh Leaves for 6 hours, 2-3 times a month.

ASTERACEAE - *Baccharis pedunculata* (Mill.) Cabr.
Pasto Miel
Shrub, Andean, 500-2500m
Use: Cysts, Wounds (clotted), Abscesses / Whole plant, dried / Topical / Heat with Agua Florida. Place heated Poultice on affected area with a cloth. Leave for 2 days as it will absorb the cyst.

ASTERACEAE - *Baccharis salicifolia* (R. & P.) Pers.
Hierba de la Plata, Chilco Hembra, Chilco Macho
Shrub, Amazonian, Andean, Coastal, 0-3500m, weed
Use: 1. Good Business, Protection, Good Fortune, Good Health / Whole plant, fresh / Topical / Mixture for Spiritual Flowering. Bathe once. **2.** Good Business, Protection, Good Fortune, Good Health / Whole plant, fresh / Seguro / Standard Seguro mixture. **3.** Allergies, Rashes, Pimples / Whole plant, fresh / Topical / Mixture for Spiritual Flowering. Bathe once. **4.** Diabetes / Whole plant, fresh / Oral / Boil 1 liter of water and 100g of plant material. Drink mixture 3 times a day for 1 month.

Baccharis tricuneata

Bidens pilosa

Chuquiragua spinosa

Chuquiragua weberbaueri

Clibadium sylvestre

Cronquistianthus lavavandulaefolius

ASTERACEAE - *Baccharis tricuneata* (L.f.) Pers.
Sigueme Sigueme
Shrub, Andean, 2000-4500m
Use: 1. Good Luck, Spiritual Flowering / Florecimiento / Flowers and Leaves, fresh / Topical / Standard Seguro mixture. Combined with a prayer invoking the name of the patient, owner of the Seguro. Spray on Tuesdays and Fridays. Spray and rub the mixture on the patient for good luck.
2. Good Luck, Spiritual Flowering / Florecimiento / Flowers and Leaves, fresh / Topical / Bathe in 50g of the following: Hierba del Lucero, Hierba del Este, Ambrocilla, Señorita, Caballero, Pega Pega, Siempre Viva, Carpintero, Waime Waime, Piri Piri (Hembra and Macho), Hierba del Buen Querer, Hierba del Oro, Hierba de la Plata, Hierba del Halago, Sigueme Sigueme, and Hierba del Negocio. Boil in 5-7 liters of water for 20 minutes, then add a bit of the following perfumes: Cariño, Dios de la Huaringa, Dios de la Felicidad, San Antonio, Macumba Pusanga, Gran Jefe, Mil Flores, Llama Plata, and Ekeko. Let it cool before bathing, 2 times (Tuesdays and Fridays only) every 3 months.

ASTERACEAE - *Bidens pilosa* L.
Amor Seco, Cadillo, Tres Esquinas, Carqueja
Herb, Amazonian, Andean, Coastal, 0-4500m, weed
Use: 1. Gallbladder, Kidney Inflammation, Inflammation (general), Kidneys, Prostate, Hair Loss, Diabetes, Liver, Blood, Heart / Whole plant, fresh or dried / Oral / 10g in 1 liter of water combined with Chacur, Unquia, Flor de Arena, Espiga de Maiz, Cola de Caballo, Guanábana, Pimpinela, and Flores de Azares. Drink 1 cup 4 times a day for 1 month. **2.** Gallbladder, Kidney Inflammation, Inflammation (general), Kidneys, Prostate, Hair Loss, Diabetes, Liver, Blood, Heart / Whole plant, fresh or dried / Topical / Same mixture can be used as bath.

ASTERACEAE - *Chuquiragua spinosa* Lessing ssp. *huamanpinta* C. Ezcurra
Chuquiragua, Huamanpinta
Shrub, Andean, 3000-4500m
Use: Inflammation, Kidneys, Prostate, Bladder, Prostate Inflammation, Sexual Impotence / Leaves, dried / Oral / Boil 5-10g in 1 liter of water for 3-5 minutes, 1-4 cups a day for 15 days or as needed. For impotence, macerate 6 plants for 8 days in wine in a 1 liter bottle. Mix with Huevos del Angelote, Pollen, Catachi, and Viril de Oso. Take as needed.

ASTERACEAE - *Chuquiragua weberbaueri* Tovar
Amaro Amaro
Shrub, Andean, 3000-4500m
Use: Cough, Bronchitis, Asthma, Liver, Mal Aire/Bad Air / Whole plant, fresh or dried / Oral / Boil 10g in 1 liter of water for 3-4 minutes with Eucalyptus, Matico, Mullaca, Muña, and Flor de Overo. Take one cup 3-4 times a day for 1 month.

ASTERACEAE - *Clibadium sylvestre* (Aubl.) Baill.
Flor de Novia
Herb, Amazonian, Andean, 0-1500m
Use: Cold, Before marriage / Flowers, Leaves and Stems, fresh or dried / Topical / 1 bundle with 20 drops of perfume in 3 liters of boiling water, 3 Baths per month.

ASTERACEAE - *Cronquistianthus lavavandulaefolius* (DC.) R.M. King & H. Rob.
Clavelillo, Espino de Hoja, Pulmonária
Herb or Shrub, Amazonian, Andean, Coastal, 500-4000m
Use: Cough, Bronchitis, Headache, Cold, Asthma, Pulmonary Disease / Flowers, Leaves and Stems, fresh or dried / Oral / Combine 10g of plant material with Matico, Zarzamora, Nogal, Salvia, Borraja, Llatama, and Vira Vira and boil in 1 liter of water. Boil the mixture for 3-4 minutes. Drink 1 liter daily for 3 months.

Cynara cardunculus

Diplostephium gynoxyoides

Diplostephium sagasteguii

Ferreyranthus verbascifolius

Flaveria bidentis

Gamochaeta americana

ASTERACEAE - *Cynara cardunculus* L.
Alcachofa (Artichoke)
Herb, Andean, Coastal, 0-3500m, introduced and cultivated
Use: 1. Diabetes, Memory Loss, Physical Weariness, Liver, Blood Purification, Mental Weariness / Stems and Leaves, fresh or dried / Oral / Boil 10g in 1 liter of water for 5 min; 1 liter a day or 3-4 glasses per day. **2.** Weight Loss / Stems and Leaves, fresh or dried / Oral / Add 1 liter of water to 1 Alcachofa and boil for 5 minutes. Drink 1 cup 3 times a day for 1 month.

ASTERACEAE - *Diplostephium gynoxyoides* Cuatrec.
Parrano
Shrub, Andean, 2500-3500m
Use: 1. Cold, Inflammation of the lungs / Flowers, fresh / Oral / Boil 10 Flowers of Parrano and 4 Leaves of Chicoria in 1/2 cup of water for 2 minutes. Patient should drink hot solution, 3 tbsp 3 times a day for 5 days. **2.** Bad Air/Mal Aire / Flowers, fresh / Topical / Crush 2kg of Parrano Leaves with 200g of Flor de Muerto, Frijol Chileno, Garlic, and Agua Cananga (red perfume). Place on affected area and cover with a piece of cloth for 3 hours.

ASTERACEAE - *Diplostephium sagasteguii* Cuatrec.
Hierba del Tigre
Shrub, Andean, 3500-4500m
Use: 1. Bad Air/Mal Aire, Fragrance, Good Luck for work, Sorcery/Daño (prevention), Undo bad things done to the victim, Strength, Evil/Maldad (cure), Evil Eye/Mal Ojo / Leaves and Stems, fresh or dried / Oral / Boil 10g of Hierba del Tigre, 10g of Hierba del Oso, and 10g of Semora Negra with 3 Leaves of Toro Cimuro and 3 Leaves of Misha Amarilla in 1/2 cup of water for 5 minutes. Very strong compound. Do not exceed the dosage: 1/8 cup once only. Drink cold. Patient should stay inside the house without any light or noise for 3 days. Should also observe a diet (no spices or seafood). **2.** Protection from evil / Leaves and Stems, fresh or dried / Topical / Bath mixture for Protection from Evil. Only once. **3.** Bad Air/Mal Aire, Fragrance, Good Luck for work, Sorcery/Daño (prevention), Undo bad things done to the victim, Strength, Evil/Maldad (cure), Evil Eye/Mal Ojo / Leaves and Stems, fresh or dried / Seguro / 2 small branches per Seguro.

ASTERACEAE - *Ferreyranthus verbascifolius* (Kunth) H. Rob. & Brettell
Tutapure Amarillo, Tutapure Amarillo (Grande)
Shrub, Andean, 1000-3000m
Use: Fright/Susto, Diarrhea in children from Susto, Daño/Sorcery / Whole plant, fresh / Topical / Mix with Timolina, 2 times a week as Limpia. Mixed with Tutapure Negro, Manzanilla Blanca, Añasquero Chico, Ruda Chingue, Conchalay, Ticra, Manzanilla, and 7 Espiritus as bath, 3 times a week. Can also be used as steam bath once a month.

ASTERACEAE - *Flaveria bidentis* (L.) Kuntze
Mata Gusano
Herb, Andean, Coastal, 0-2500m, weed
Use: Cough, Bronchitis / Flowers, Leaves and Stems, fresh or dried / Oral / Boil in 1 liter of water, then add 10g of Mata Gusano. Drink 3-4 times a day for 1-2 weeks, or as needed.

ASTERACEAE - *Gamochaeta americana* (Mill.) Wedd.
Lechuguilla
Herb, Andean, Coastal, 1000-4500m, weed
Use: Diabetes, Nerves / Whole plant, fresh or dried / Oral / Boil 1/2 liter of water with 10g of Lechuguilla. Patient should drink lukewarm solution, 1 glass 2-3 times a day for 1 month.

Lactuca sativa

Loricaria ferruginea

Loricaria thyrsoidea

Matricaria chamomilla

Matricaria recutita

Mikania leiostachya

ASTERACEAE - *Lactuca sativa* L.
Lechuga (Lettuce)
Herb, Amazonian, Andean, Coastal, 0-3500m, introduced and cultivated
Use: Nerves, Blood purification, Cleans Toxins from blood / Root and Stems, fresh / Oral / Add 100g of the plant material to 100g of Beets and 1/2 liter of water and heat. After heating, let mixture cool. Drink 1 cup 2-3 times a day for 1 month.

ASTERACEAE - *Loricaria ferruginea* (Ruiz & Pav.) Wedd.
Palmerilla, Palmita, Pata de Gallina, Palmera, Trensilla, Palmilla, Patita de Gallo, Palmia Pina, Palmera Blanca, Destrensilla
Shrub, Andean, 3000-4500m
Use: 1. Spiritual Flowering / Florecimiento, Menstrual Delay, Blood Circulation / Leaves and Stems, fresh / Topical / Alternative mixture for Spiritual Flowering. Apply 3 times a day for 1 week; only once for Spiritual Flowering/Florecimiento. **2.** Protection, Good Health, Good Fortune, Good Business, Fragrance, Success, Safe Travel, Sociability, Good Relations with others / Leaves and Stems, fresh / Seguro / Standard Seguro mixture.

ASTERACEAE - *Loricaria thyrsoidea* (Cuatrec.) Dillon & Sagástegui
Palmilla Ancha, Palmilla Verde, Palma Bendita
Shrub, Andean, 3500-4500m
Use: 1. Business / Whole plant, fresh / Seguro / 7 small plants per Seguro combined with strong magical herbs. **2.** To cast out Evil Spirits / Whole plant, fresh / Topical / Boil 15-20 minutes, 10-20g per 12 liters of water. Bath: 2-3 times a month.

ASTERACEAE - *Matricaria chamomilla* L.
Manzanilla (Camomille)
Herb, Andean, Coastal, 0-4000m, introduced and cultivated
Use: 1. Love-sickness, Nerves, Insomnia, Inflammation of wounds, Colic, Stomachache, Bronchitis / Whole plant, fresh or dried / Oral / Boil water first. Add 10g of Manzanilla per cup, 3 cups a day for 1 week. **2.** Inflammation, Colic, Inflammation of the Vagina, Injuries, Wounds (open), Wounds (closed) / Whole plant, fresh or dried / Topical / Boil water first. Add 10g Manzanilla per cup. Do not mix with other herbs. Rub solution over the womb or inflamed area as needed. For vaginal inflammation, squat over the steam 2-3 times a day, every other day. Alternatively, boil Manzanilla, then place on a cloth. May also boil a Manzanilla tea bag. Place cloth with herbs or tea bag on affected area for 3-4 minutes or until the tea bag or cloth is cool. Heat again and repeat the process 3-4 times a day for no more than 2 days.

ASTERACEAE - *Matricaria recutita* L.
Manzanillón, Agua de la Banda, Manzanilla Blanca, Manzanilla Amarga, Manzanilla
Herb, Andean, 2000-4500m, introduced and cultivated
Use: 1. Safe Travel, Sociability, Good Relations with others / Whole plant, fresh / Seguro / 3 Stems per flask. **2.** Fright/Susto, Infection from Wounds, Vaginal Cleansing / Whole plant, fresh / Topical / Boil the whole plant for 3-5 minutes. Mix entire plant with Hierba del Susto and Ajenco after combining with 2 tsps of Vinegar. Pour the mixture in a bathtub and sit in it. Rub 2-4 times a month. Boil 1 dried bundle in 1 liter of water for 5 min to wash wounds. **3.** Blood Purification, Menstrual Colic / Whole plant, fresh / Oral / Boil water. Add 10g of Manzanillón to 1 cup of hot water. Manzanilla, Toronjil and Pimpinela may be added. Take 1 small cup 3 times a day for 1 month. Drink lukewarm. **4.** Infection from Wounds / Whole plant, fresh / Topical / Chop fresh herb to get extract and mix with odorless Vaseline. Wash with Llantén, then apply ointment as needed.

ASTERACEAE - *Mikania leiostachya* Benth.
Enredadera
Liana, Amazonian, Andean, 0-2000m
Use: Daño/Sorcery, Daño de Brebaje/Drink Sorcery / Leaves, dried / Topical / 1 handful in 3 liters of boiled water combined with 10 g each of Huaminga, Chilca, Hierba del Susto and Agua del Susto. Take 2 baths per week.

Monactis flaverioides

Munnozia lyrata

Onoseris odorata

Oritrophium peruvianum

Paranephelius uniflorus

Perezia multiflora

ASTERACEAE - *Monactis flaverioides* Kunth
Hierba del Susto (Amarillo), Malva, Mocura, Hierba del Susto, Hierba Susto
Shrub, Amazonian, Andean, Coastal, 0-3000m
Use: 1. Bad Air/Mal Aire, Prostate, Soul Loss/ Susto (pérdida del alma), Vaginal Cleansing, Business, Bad Luck/Mala Suerte (cure), Fright in Children/Susto de Niños / Stems and Leaves, fresh / Topical / 7 Stems with their Leaves boiled in 3-5 liters of water for 20 min and combined with 10 g each of Agua del Susto, Ajenco and Llatama. Bath: 2-3 times a week at 7, 9 and 11PM.
2. Bad Air/Mal Aire, Prostate, Soul Loss/ Susto (pérdida del alma), Vaginal Cleansing, Business, Bad Luck/Mala Suerte (cure), Fright in Children/Susto de Niños / Stems and Leaves, fresh / Oral / Mix with Toronjil and Pimpinela and boil for 3-5 minutes, 1 liter daily for 7 days.

ASTERACEAE - *Munnozia lyrata* (A. Gray.) H. Rob. & Brettell
Canillahuanga
Herb, Andean, 2000-4000m
Use: Fright/Susto, Bad Air/Mal Aire / Whole plant, fresh or dried / Topical / Bath: 20g total per 5 liters of water, boiled for 20 minutes with Hierba del Susto, Añasquero, Cutiquero, Hierba del Ave, and Ishpingo, 3 times a week.

ASTERACEAE - *Onoseris odorata* (D. Don) Hooker & Arnott
Hierba de la Reina
Herb, Amazonian, Andean, Coastal, 0-4000m
Use: Heart, Nerves / Whole plant, fresh or dried / Oral / 10g in 1 liter of boiling water, 3 cups a day.

ASTERACEAE - *Oritrophium peruvianum* (Lam.) Cuatrec.
Huamanripa, China Linda, Wiña Wiña, Vira Vira, Hórnamo, Hierba del Sol, Maguanmarica, Hierba del Lucero
Herb, Andean, 3500-4500m
Use: 1. Asthma, Bronchitis, Pneumonia / Whole plant, fresh or dried / Oral / Add 10g of plant material to 1 liter of water and boil for 3 minutes, 3 cups a day, as needed. Drink lukewarm. **2.** Fragrance, Attract Lovers, Spiritual Flowering / Whole plant, fresh or dried / Topical / Standard mixture for Spiritual Flowering. **3.** Illuminating your path and destiny, Good Business, Protection, Good Fortune, Good Health / Whole plant, fresh or dried / Seguro / Standard Seguro mixture.

ASTERACEAE - *Paranephelius uniflorus* Poepp. & Endl.
Pacha Rosa, Carapa de Chancho
Herb, Andean, 3000-4500m
Use: Inflammation of the Ovaries, Uterus, Inflammation (internal female organs), Stones, Inflammation / Whole plant, fresh or dried / Oral / 5g each in 1 liter of boiling water mixed with Flor Blanca, Purenrosa, Flor de Arena, Manayupa, Sauco, Cola de Caballo, and Pie de Perro. Other anti-inflammatory plants can be mixed in as well. Addition of these other plants is optional. Take 3-4 times a day for one month.

ASTERACEAE - *Perezia multiflora* (Humb. & Bonpl.) Lessing
Corzonera, Escorcionera, Escorzonera
Herb, Andean, 3500-4500m
Use: Nerves, Cough, Bronchitis, Asthma, Sharp pain in the body / Whole plant, fresh or dried / Oral / Boil 1 liter of water, then add 10g each of Escorcionera combined with Matico, Eucalyptus, Veronica, Vira Vira, Nogal, Huamanripa, Tilo, and Zarzamora. Drink 3 cups a day for 15 days. Patient should drink cold solution.

Perezia pungens

Picrosia longifolia

Pluchea absinthioides

Porophyllum ruderale

Pseudogynoxys cordifolia

Schkuhria pinnata

ASTERACEAE - *Perezia pungens* (Kunth) Cass.
Lengua de Vaca
Herb, Andean, 3000-4500m
Use: Wound Infections. Prevents peeling of skin after sunburn, Twisting or Fractures from Sorcery / Leaves, fresh / Topical / 1 bundle of Leaves makes a cream. Apply 2 times a week.

ASTERACEAE - *Picrosia longifolia* D. Don
Achicoria, Chicoria
Herb, Coastal, 0-500m
Use: Liver, Blood, Hepatitis, Gallbladder, Purification of the Blood, Bronchitis, Pneumonia / Whole plant, fresh / Oral / Boil 10-50g of Chicoria, Verbena, Canchalagua and Chochocon in 1 liter of water. Drink 1 liter daily for 15-30 days. Alternatively, chop and extract juice of 200g of fresh material. Drink 1 glass daily, no longer than a week. Overdosing can harm vision.

ASTERACEAE - *Pluchea absinthioides* Hook. & Arn.) H. Rob. & Cuatr.
Pata de Gallina
Shrub, Andean, 2000-2500m
Use: Protection of Job and Home, Protection (general) / Whole plant, fresh or dried / Seguro / Mix in a bottle 10g of Valeriana Estrella, Señorita, Carpintero, Chupa Flor, Hierba de la Coqueta, Oro, Dolar. Add Agua Florida, Ramillete de Novia, Tabú, Lime Juice, Agua Bendita, and Sugar. 1 bottle to use on a regular basis.

ASTERACEAE - *Porophyllum ruderale* (Jacq.) Cass.
Hierba Gallinazo, Hierba del Gallinazo
Shrub, Amazonian, Andean, Coastal, 0-2000m, weed
Use: **1.** Clean the energy of the home / Whole plant, dried / Incense / Burn with Llatama, Ajosquiro and Añasquero Chico, 5g of each herb, 2 times a month. **2.** Daño/Sorcery, Fright/Susto / Whole plant, dried / Topical / Combine 5g of Llantén, Ajosquiro, Hierba Gallinazo, Hierba del Romero, Flor del Muerto, Eucalyptus, Floripondio Flowers, Retama and Añasquero Chico to 3 liters of water. Bathe 2 times a month, Tuesday and Friday only. Rub body with herbs. Rinse with the water. Do not dry with a towel. Air dry.

ASTERACEAE - *Pseudogynoxys cordifolia* (Cass.) Cabrera
San Juan
Vine, Amazonian, Andean, Coastal, 0-2500m
Use: Fright/Susto, Bad Air/Mal Aire / Whole plant, fresh or dried / Topical / Add 10g of San Juan, Eucalyptus, Chancas del Muerto, and Romero to 2 liters of water. Boil the mixture for 3 minutes. Wash the patient in the lukewarm mixture in the evening. Advise the patient not to leave the house afterwards. Bathe every 2 days.

ASTERACEAE - *Schkuhria pinnata* (Lam.) Kuntze
Canchalagua, Canchalagua Chica
Herb, Andean, 1000-3000m, weed
Use: Blood Cleansing, Liver, Gallbladder, Bad Breath, Diabetes, Menstrual delay, Allergies, Menstruation, Blood Irrigation, Inflammation of the Urinary Tract. / Whole plant, fresh / Oral / Boil 20g total in 1 liter of water for 3-10 minutes mixed with Ortiga, Lancetilla, Culantrillo, Panisara, Purenrosa, Boldo, Berro, Flor Blanca, and Canchalagua. Take 3 times a day (1 liter) for 1 month. Blood Purification: Brew must sit out overnight before drinking.

Senecio canescens

Senecio comosus

Senecio genisianus

Senecio hypsiandinus

Senecio otuscensis

Senecio tephrosioides

ASTERACEAE - *Senecio canescens* (H.B.K.) Cuatrecasas
Vira Vira, Oreja de Conejo
Herb, Andean, 3500-4500m
Use: 1. Bronchitis, Asthma, Cough, Nerves / Whole plant, fresh / Oral / Boil 10g each of diced herb in 1 liter of water combined with Borraja, Eucalyptus, Escorcionera, Borraja, Cerraja, Polen de Hierbas, Manzanilla, Toronjil, Congona, Poleo, Claveles, Juan Alonso, Espina de Hoja, and Camphor. Drink 3 cups a day for 1 month. **2.** Bronchitis, Asthma, Cough, Nerves / Whole plant, fresh / Topical / Use same mixture for steam baths and inhalation.

ASTERACEAE - *Senecio comosus* Sch.-Bip.
Hórnamo Leon Amarillo
Herb, Andean, 3500-4500m
Use: 1. Bad Air/Mal Aire, Inflammation (general), Vision Enhancement / Leaves and Stems, fresh or dried / Oral / Boil 3 pieces (3 inches each) of Hórnamo León Amarillo and Hórnamo León Verde, 2 slices of San Pedro (6 ribs and 7 ribs) as well as 3 pieces (3 inches each) of Condor Purga in 9 liters of water for 1 hour at a low temperature. Drink cold, 1 small glass once only. **2.** Bad Air/Mal Aire, Inflammation (general), Vision Enhancement / Leaves and Stems, fresh or dried / Oral / Boil 2 leaves of Misha Morada, 1 leaf of Misha Amarilla, 1 leaf of Misha Blanca, 1 leaf of Misha Rosada, 1g of Toromaique and 1g of Toro Misha in 1/2 cup of water for 5 minutes. Drink cold, 1/8 of a small glass. Patient must stay in a dark room for 3 days while maintaining a diet without spices or seafood. Patient should rest for 3 more days afterwards.

ASTERACEAE - *Senecio genisianus* Cuatr.
Tutapure Blanco
Herb, Andean, 4000-5000m
Use: Wounds (cleansing), Rabies, Animal Bites / Leaves and Stems, dried / Topical / 1 handful per 3 liters of boiled water. Can combine (1 handful each) with Chuque, Huaminga, Chinque, Manzanilla de Cerro, Vinegar and 7 Espiritus. One bath a week, also for Limpias.

ASTERACEAE - *Senecio hypsiandinus* Cuatr.
Hórnamo Blanco
Herb, Andean, 4000-5000m
Use: Fragrance, Good Luck / Whole plant, fresh / Seguro / A few Stems per flask.

ASTERACEAE - *Senecio otuscensis* Cabrera
Árnica
Shrub, Andean, 2500-3000m
Use: Inflammation, Rheumatism, High Fever / Leaves and Stems, fresh / Topical / In 1/2 liter of water boil 100g of Árnica for 10 minutes. Wet a piece of cloth in the warm brew and place cloth on affected area for a few seconds. Repeat over and over again until body temperature is lowered. Alternatively, crush 200g and add 8 drops of alcohol to be warmed-up in a pot over the fire. Place Poultice mixture on top of the affected area, then cover with a piece of cloth covered in turn with a piece of plastic. Apply 2 times a week as needed.

ASTERACEAE - *Senecio tephrosioides* Turcz.
Huamanripa, Genciana
Herb, Andean, 3000-4500m
Use: Bronchitis, Asthma, Pneumonia / Whole plant, fresh / Oral / Boil 1 cup of water, then add 10g of Huamanripa combined with Veronica, Vira Vira and Brochamelia. Drink 3 cups a day for 15 days.

Smallanthus sonchifolius

Sonchus oleraceus

Spilanthes leiocarpa

Tagetes elliptica

Tagetes erecta

Tagetes filifolia

ASTERACEAE - *Smallanthus sonchifolius* (Poepp. & Endl) H. Rob.
Hojas de Yacón, Llacón
Herb, Andean, 2500-3500m, cultivated
Use: Diabetes, Kidneys, Inflammation of the Prostate, Cholesterol / Leaves, dried / Oral / Boil 5g in 1 liter of water. Drink 3 times a day, 1 liter in total.

ASTERACEAE - *Sonchus oleraceus* L.
Cerraja, Serraja, Zeraja
Shrub, Amazonian, Andean, Coastal, 0-4500m, weed, introduced
Use: 1. Anger, High Blood Pressure / Whole plant, fresh / Oral / Tea: 5g each of Cerraja, Colores, Lancetilla, Contrahierba in 1 liter of boiling water. Drink 3 times a day. **2.** Embarassment, Shame, Hangover, Negative Energy, Anger / Whole plant, fresh / Oral / 1 cup with the whole plant (especially Flowers) cut with scissors together with 3 drops of lime, a pinch of salt, and a shot of pisco. Let mixture sit for 3 minutes and drink 1 cup 2 times a day for 1 day until condition passes. **3.** To calm strong character / Whole plant, fresh / Oral / 1 handful of herb boiled with 1/2 liter of water, 3 cups a day before eating. **4.** Hangover / Whole plant, fresh / Oral / Take 1 whole plant, crush and drink extract, once only

ASTERACEAE - *Spilanthes leiocarpa* DC.
Turre
Herb, Andean, Coastal, 0-1000m, weed
Use: 1. Toothache, Anesthetic / Flowers, fresh / Topical / Poultice: Crush and grind the plant material. Place plant material on affected area. Apply 2 times a day as needed. **2.** Skin rashes / Leaves and Stems, fresh / Topical / Patient should bathe in the solution when it is cold. Do not rinse it off. Patient must be air dried. 2 times a week until rash is gone.

ASTERACEAE - *Tagetes elliptica* Sm.
Culantrillo Serrano
Herb, Andean, 3000-4000m
Use: Colds, Bronchitis, Congestion / Whole plant, fresh or dried / Oral / 5g of the plant in 1 cup of water boiled for 5 minutes. Drink cold, 1/4 cup a day for 8 days.

ASTERACEAE - *Tagetes erecta* L.
Flor del Muerto, Clavel Chino, Flor de Muerto
Herb, Amazonian, Andean, 0-3500m, weed, cultivated
Use: 1. Fright/Susto, Colic of the Stomach, Bad Air/Mal Aire / Flowers, Stems and Leaves, fresh or dried / Topical / Macerated in perfume along with 100g of Parrano, 100g of ground Frijol Chileno, 100g of ground Ajo, 5g of Chili powder, 5g of Black Pepper, 5g of Oregano, 5g of Flor de Chocho, 5g of Retama, 5g of Ruda (Hembra and Macho) and 5g of Agua del Susto. Poultice: Put a portion on feet and cover with a cloth. Can also be used on the stomach. Alternatively, the mixture can be used as a Bath, 3 times a week, Tuesday, Friday, Tuesday. **2.** Cough, Nerves, Inflammation (general) / Flowers, Stems and Leaves, fresh or dried / Oral / Take 3-4 Flowers and boil in 1 liter of water along with 10g of a mixture of Toronjil, Pimpinela, Poleo and Manzanilla. Drink 3-4 glasses a day for 1 month. **3.** Fright of Death/ Susto de la Muerte / Flowers, Stems and Leaves, fresh or dried / Topical / Limpia: Boil 5g in 3 liters of water mixed with 5g each of Ajenco, Ruda Hembra, Hierba del Susto, Manzanilla Blanca and Timolina, 2 times a month.

ASTERACEAE - *Tagetes filifolia* Lag.
Anís, Anís Serrano
Herb, Andean, 2500-3500m
Use: Severe Colic, Stomach, Stomach Pain, Diarrhea / Whole plant, fresh or dried / Oral / Boil 10g each of Anís, Poleo, Manzanilla, Muña or Chancas de Comida, and Hinojo in 1 liter of boiling water. Drink 3 cups daily for 1 week to 1 month.

Taraxacum officinale

Tessaria integrifolia

Trixis cacalioides

Weddelia latifolia

Werneria nubigena

Werneria pygmaea

ASTERACEAE - *Taraxacum officinale* Wiggers
Diente de León, Amargón, Hierba del León
Herb, Amazonian, Andean, Coastal, 0-4500m, weed, introduced
Use: 1. Liver, Stomach, Inflammation (internal), Ovaries, Bad Air/Mal Aire, Protection against Evil / Whole plant, fresh / Topical / Prepare 200g of Leaves in a flask of 7 Espiritus. Poultice: Apply 2 times a month. **2.** Liver, Stomach, Inflammation (internal), Ovaries, Bad Air/Mal Aire, Protection against Evil / Whole plant, fresh / Oral / Boil 20g of Leaves in 2 liters of water with 20g total of Chacur, Pie de Perro, Cola de Caballo, Linaza, Malva and Amor Seco. Add a piece (4 inches) of Bejuco de Contra Aire and 2 Spikes of Palmerilla. Boil for 3 minutes. Drink 1 cup 4 times a day for 1 month. **3.** Protection from Evil / Whole plant, fresh / Topical / Bath in mixture for Protection from Evil. Once only.

ASTERACEAE - *Tessaria integrifolia* R. & P.
Pájaro Bobo
Shrub, Amazonian, Andean, Coastal, 0-2500m, weed
Use: Liver, Kidneys, Gallbladder, Inflammation (general), Fever, Bad Breath / Flowers and Leaves, fresh / Oral / Boil 10g of Pájaro Bobo in 1 liter of water combined with Cola de Caballo, Verbena, Chacur, Paja Blanca and Espiga de Maiz. Drink 3-4 times a day for 15 days. Patient should drink hot solution for most ailments and cold solution for bad breath.

ASTERACEAE - *Trixis cacalioides* Kunth
Añasquero Chico
Shrub, Amazonian, Andean, Coastal, 0-2500m
Use: 1. Rid the house of negative energy / Whole plant, fresh or dried / Incense / Burn 2 times a month. **2.** Daño/Sorcery, Fright/Susto, Bad Air/Mal Aire / Whole plant, fresh or dried / Topical / Mix with Añasquero Grande, Ruda Hembra, Ruda Macho, Ajenco, Timolina, Ishpinguillo, Chuque and 7 Espiritus. Take 2 baths a week or use as Limpia once a week.

ASTERACEAE - *Weddelia latifolia* DC.
Chulgan, Cuchalman
Shrub, Amazonian, Andean, Coastal, 0-2000m
Use: 1. Fever / Whole plant, fresh / Oral / Boil 10g of Chulgan with 1 liter of water. Patient should drink the solution at room temperature, once only. **2.** Fever / Whole plant, fresh / Oral / Boil 1 small bundle of Chulgan with 2 liters of water. Do not mix with other herbs. Patient should drink lukewarm solution, once only.

ASTERACEAE - *Werneria nubigena* Kunth
Hierba de la Señorita
Herb, Andean, 2500-4000m
Use: Inducement of love in men and women, Making the feelings of men and women clearer / Flowers, Leaves and Stems, fresh / Seguro / 3 Flowers or Stems per flask.

ASTERACEAE - *Werneria pygmaea* Gillies ex Hook. & Arn
Hierba del Halago
Herb, Andean, 3500-5000m
Use: 1. Good Business, Protection, Good Fortune, Good Health / Whole Fruit, fresh / Seguro / Standard Seguro mixture. **2.** Spiritual Flowering / Whole Fruit, fresh / Topical / Standard mixture for Spiritual Flowering.

Werneria villosa

Corynaea crassa

Berberis buceronis

Alnus acuminata

Crescentia cujete

Cydista aequinoctialis

ASTERACEAE - *Werneria villosa* A. Gray
Hierba del Oro
Herb, Andean, 3000-4000m
Use: 1. So that all goes well in the life and Home of a person, Good Luck, Large Enterprises, Personal Life, Good Business, Protection, Good Fortune, Good Health / Leaves and Stems, fresh / Seguro / Standard Seguro mixture. **2.** So that all goes well in the life and Home of a person, Good Luck, Large Enterprises, Personal Life, Good Business, Protection, Good Fortune, Good Health / Leaves and Stems, fresh / Topical / Alternative mixture for Spiritual Flowering. After boiling, add a bottle of your favorite perfume. Rub the entire body with all the herbs, then rinse with water and air dry. Do not use soap or a towel.

BALANOPHORACEAE - *Corynaea crassa* Hook. f.
Huanarpo (Hembra and Macho)
Parasitic herb, Andean, 1500-2500m
Use: Fertility, Sexual Potency, Male Impotence, Tension / Tuber/Root, fresh / Oral / To 1 bottle of wine add 10g of Huanarpo. Then add 10 g each of Huevo del Angelote, Bee Pollen, Honey, Pacra, Palo Sangre, Palo Huaco, Chuchuhuasi, Cascarilla, and Para-Para. Drink 3 cups a day for 3-6 months. Take a small cup before intercourse. Can refill the wine bottle once more with the same herbs and it will be stronger. For a man, use Macho for a woman, use Hembra. Drink 3 cups a day for 3-6 months.

BERBERIDACEAE - *Berberis buceronis* J.F. Macbride
Palo Amarillo
Shrub, Andean, 2000-2500m
Use: Liver, Hepatitis / Wood and Bark, dried / Oral / Boil 2g of the Bark in 1 liter of water for 3 minutes with a total of 10g of Amor Seco and Cola de Caballo. Drink lukewarm with 3 drops of Lime Juice, 3 cups a day for 1 month.

BETULACEAE - *Alnus acuminata* Kunth
Aliso Blanco (Liso), Aliso Colorado (Arrugado)
Tree, Andean, Coastal, 0-4000m, cultivated
Use: 1. Sealing Wounds, Rashes, Skin Irritations, Arthritis / Bark, fresh / Topical / Macerate 1kg of Aliso Colorado Bark mixed in 4 liters of alcohol. Apply to wounds. Do not Ingest! Apply until wound seals, 3 times a day. **2.** Arthritis, Cold, Colic of the Stomach, Colic of the Intestine / Bark, fresh / Oral / Boil for 10 minutes, 2 tbsp per cup to obtain the extract. Take 5g every 4 hours. **3.** Bone Pain, Arthritis / Bark, fresh / Topical / Soak Bark in water and use as Bath or grind the Leaves and mix with odorless Vaseline. Bathe 3 times a month or rub daily on the patient until symptoms improve.

BIGNONIACEAE - *Crescentia cujete* L.
Higuerón
Tree, Amazonian, 0-500m, cultivated
Use: Healing of belly button after birth / Latex from Leaf, fresh / Topical / With the milk of Higuerón form a ball with the Latex. Apply with white cotton to the bellybutton and tie it down with pressure. Leave it on for 3 weeks.

BIGNONIACEAE - *Cydista aequinoctialis* (L.) Miers
Bejuco Amarillo
Liana, Amazonian, Andean, 0-1000m
Use: Sorcery/Daño, Internal Inflammation from Sorcery, Bruises from Sorcery / Flowers, Leaves and Stems, fresh / Topical / Boil 1 handful in 3 liters of water for 5 minutes, 2 times a month.

Jacaranda acutifolia

Tynanthus polyanthus

Bixa orellana

Borago officinalis

Cordia alliodora

Cordia lutea

BIGNONIACEAE - *Jacaranda acutifolia* Humb. & Bonpl.
Arabisca, Yarabisca
Tree, Andean, 1000-2500m
Use: Cough, Bronchitis, Asthma, Phlegm / Leaves and Stems, fresh or dried / Oral / Boil 10g in 1 liter of water for 2-3 min. Drink 3 cups a day as needed.

BIGNONIACEAE - *Tynanthus polyanthus* (Bureau) Sandwith
Clavo Huasca
Liana, Amazonian, Andean, 0-1000m
Use: Bad Air/Mal Aire, Hallucinogen, Enhance the Vision of the Shaman / Leaves and Stems, fresh / Oral / Blend 100g of plant material with 1/4 glass of water and drain. Drink cold. Use during ritual ceremonies, 5g per ritual.

BIXACEAE - *Bixa orellana* L.
Achiote, Hoja de Achiote
Tree, Amazonian, Andean, 0-1000m, weed and cultivated
Use: Kidney Inflammation, Prostate, Food Coloring, Bronchitis, Hemorrhages, Pulmonary System, Urinary Infections / Seeds and Leaves, fresh or dried / Oral / Chop 3 seeds and eat as needed. Alternatively, boil and add 10g of plant material to 10g of Uña de Gato in 1 liter of water. Boil the mixture for 3-4 minutes and mix with Chante because the plant is cold. Drink 1 liter a day for 1 week. Patient should drink warm solution. If possible, use Leaves.The Seeds are less powerful.

BORAGINACEAE - *Borago officinalis* L.
Borraja (Borage)
Herb, Andean, Coastal, 0-3500m, introduced and cultivated
Use: Bronchitis, Lungs, Blood, Lose Weight, Anxiety, Depression, Heart, Nerves, Insomnia, Cough, Cold, Bruises / Whole plant, fresh or dried / Oral / Place 10g total of the herb in 1 liter of boiling water (boiled for 3-5 minutes) combined with 10g of Vira Vira. Drink 3 times a day or 1 liter a day for as long as needed.

BORAGINACEAE - *Cordia alliodora* (R. & P.) Oken
Ajosquiro, Ajos Quiro, Ajo Sacha
Tree, Amazonian, Andean, 0-1500m
Use: 1. Daño/Sorcery, Fright/Susto, Dispelling negative energy from the house / Bark and Stems, dried / Topical / Combine 5g each of Llatama, Ajosquiro, Añasquero Grande, Llatama, Hierba del Gallinazo, Añasquero Chico, and Ruda Macho in 3 liters of water for 2 Baths a month. Can also be used as a steam bath. **2.** Bronchitis / Bark and Stems, dried / Oral / Add 1 bottle of Wine to 10g of plant material and 20g total of Chuchuwasi, Cascarilla, Honey, Pollen, and Tutuma. Let the mixture sit for 1 week. Drink the mixture. Patient should not leave the house while taking treatment. Adults take 1 small cup. Children take 1 tsp. Patients take the medication 3-4 times a day until the bottle is finished.

BORAGINACEAE - *Cordia lutea* Lam.
Overo, Flor de Overo, Overal
Shrub or tree, Andean, Coastal, 0-1500m, weed
Use: Liver, Bladder, Hepatitis, Inflammation of the Kidneys, Prostate Inflammation / Flowers, fresh or dried / Oral / Place 5g in 1 liter of water with Llantén and Boldo. Boil for 5 minutes. The plants need to be gathered January-February. Drink 4 cups a day for 1 month, after meals. After drinking the beverage, eat a lemon candy. Patient must limit physical activity until well rested. Latex of the Fruit is used as paper glue.

Heliotropium curassavicum

Tiquilia paronychioides

Brassica oleracea

Brassica rapa

Capsella bursa-pastoris

Lepidium virginicum

BORAGINACEAE - *Heliotropium curassavicum* L.
Alacrán, Alacrancillo
Herb, Andean, Coastal, 0-2500m, weed
Use: Spiritual Flowering / Whole plant, fresh or dried / Topical / Boil 10g of Alacrán with 10g total of Hierba de la Plata, Hierba de la Justicia, and Yellow, Red and White Roses in 2-3 liters of water. Bathe 3 times (Tuesday, Friday and the following Tuesday).

BORAGINACEAE - *Tiquilia paronychioides* (Phil.) Rich.
Flor de Arena, Paja de Lagartija, Mano de Ratón
Herb, Andean, Coastal, 0-1500m, weed
Use: Inflammation, Kidney Inflammation, Ovarian Inflammation, Gallbladder Stones, Prostate Inflammation, Gallbladder, Urinary Infections / Flowers, fresh or dried / Oral / Combine 10-100g of plant material with 10g total of Malva, Espiga de Maiz, Cola de Caballo, Contrahierba, Flor Blanca, Cadillo, Berros, Chante, Achiote, Lancetilla, and Pomanpara. Boil 3-5 minutes in 1 liter of water. Drink 3-4 times a day, 1 liter daily for 2 weeks to 1 month.

BRASSICACEAE - *Brassica oleracea* L.
Col, Repollo (Cabbage)
Herb, Andean, 2500-3500m, introduced and cultivated
Use: Gallbladder Stones / Leaves, fresh / Oral / Combine 3-4 Leaves of Cabbage in 1 liter of water with a couple of drops of Olive Oil. Drink lukewarm, 1 cup 3 times a day for 1 week.

BRASSICACEAE - *Brassica rapa* L.
Nabo (Raddish)
Herb, Andean, 2000-4000m, introduced and cultivated
Use: **1.** Throat Infection and Inflammation / Root, fresh / Topical / Grind tuber and drain to extract the juice. Gargle quickly with the juice 3 times a day for 2-3 days. **2.** Kidney Inflammation, Ovaries / Root, fresh / Topical / Grind 2 big tubers and place on the affected area. Cover with a piece of cloth for 5 minutes, 3-4 times a day for 2 days.

BRASSICACEAE - *Capsella bursa-pastoris* (L.) Medic.
Bolsita del Pastor, Hierba del Pastor, Bolsa de Pastor (Sheppard's purse)
Herb, Andean, Coastal, 0-4500m, weed, introduced
Use: Kidneys, Prostate, Inflammation (general), Inflammation (internal), Liver, Gallbladder, Stomach Infection, Urinary tract / Whole plant, fresh or dried / Oral / Combine 10-30g total in 1 liter of water mixed with Chacur, Verbena, Espiga de Maiz, Flor Blanca, Cola de Caballo, Flor de Arena, Pasuchaca, Corpus Way, Cola de Caballo, and Arenilla. Drink 4 cups a day for 1 month as needed.

BRASSICACEAE - *Lepidium virginicum* L.
Maipa
Herb, Andean, Coastal, 0-1500m, weed, introduced
Use: Pockmarks (Facial), Sunspots, Malnutrition Blemishes, Skin Blemishes (Facial), Wounds / Whole plant, fresh / Topical / Boil 1 cup of water and mix with 2 small branches, or 1 small branch for a blemish on the skin. Wash the wound with water in the morning, afternoon, and evening. Wash face 3 times a day.

Raphanus sativus

Rorippa nasturtium-aquaticum

Ananas comosus

Puya hamata

Puya weberbaueri

Tillandsia cacticola

BRASSICACEAE - *Raphanus sativus* L.
Rabanito (Raddish)
Herb, Andean, 2000-3500m, introduced and cultivated
Use: 1. Bronchitis / Tuber, fresh / Oral / To 1/4kg of sugar add 1/2kg of Rabanito cut in pieces. Boil with 1 Green Onion with no water. The syrup becomes a drink for the patient. Drink 5g every 6 hours for 1 month. **2.** Blood Detoxification, Liver Cleansing, Face Blemish Erasure / Tuber, fresh / Oral / Blend 50g of Rabanito. Drink cold in the morning while fasting, 1 glass once a day for 15 days. Drink while fasting.

BRASSICACEAE - *Rorippa nasturtium-aquaticum* (L.) Hayek
Berros (Watercress)
Herb, Andean, Coastal, 0-3500m, weed, introduced
Use: Liver, Urine Retention, Bronchitis, Kidneys, Inflammation of the Liver, Inflammation of the Kidneys, Anemia / Whole plant except root, fresh or dried / Oral / Drink fresh as needed or grind and drink the juice with Alfalfa. Make a soup with the nape of the neck of the sheep and boil. Add potatoes and vegetables. Alternatively boil 1 liter of water with Berros plus 10g total of Malva, Pie de Perro, Unquia, Amor Seco, Chacur, Paja Blanca, Flor de Arena, and Purenrosa. Boil for 3-4 minutes. Drink 3-4 times a day for 1 month.

BROMELIACEAE - *Ananas comosus* (L.) Merrill
Piña (Pineapple)
Herb, Amazonian, Andean, 0-1500m, cultivated
Use: Burn fat, Lose weight / Fruit peel and Fruit, fresh / Oral / Whole pineapple peel per 1 liter of water boiled for 3-4 minutes. Drink hot, 1 cup 3 times a day as needed. Also drink 1 glass of fresh juice daily.

BROMELIACEAE - *Puya hamata* L.B. Sm.
Hierba del Carnero, Hierba de Borrego
Herb, Andean, 3000-4000m
Use: 1. Making a man stupid, Making a man obedient like a sheep, Cleansing, Controlling a violent person, Dominating a drunkard, Tumors, Infections / Hairy part of the Seeds, dried / Oral / Combine 1 cup of water and 5g of the plant (the hairy part of the seed is the most important) and boil for 3 minutes. Drink 1 cup twice a day 3-4 times a week. This Seguro is used to get the patient under control, usually if the patient is being violent or out of control because of drunkenness. **2.** Making a man stupid, Making a man obedient like a sheep, Cleansing, Controlling a violent person, Dominating a drunkard, Tumors, Infections / Hairy part of the Seeds, dried / Topical / Same mixture can be applied as a Poultice.

BROMELIACEAE - *Puya weberbaueri* Mez.
Ticta, Tifta
Herb, Andean, 2000-4000m
Use: Bad Air/Mal Aire, Wounds, Any illness involving wounds / Whole plant, fresh or dried / Topical / Boil 15g of Ticta and 10g of Hierba Santa in 3 liters of water. Boil the mixture for 3-4 minutes. Bathe the patient in the mixture. Patient can bathe on any day. Bathe once a week for 1 month.

BROMELIACEAE - *Tillandsia cacticola* L.B. Sm.
Palmera, Siempre Viva, Palma Bendita, Siempreviva (lilac)
Herb, Andean, Coastal, 0-3000m
Use: 1. Fright/Susto, Heart, Gases, Nerves, Anxiety, Heavy Air, Good Luck, Fright of Death/ Susto de la muerte, Spiritual Flowering, Good Business, Protection, Good Luck, Good Health / Leaves and Stems, fresh / Topical / Place 50g in a brazier combined with Romero, Palo Santo, Alucema, Incense, Saumerio and Myrrh. As Bath: Alternative mixture for Spiritual Flowering. Steam Bath as needed or bath once a day for 15-30 days. **2.** Fright/Susto, Heart, Gases, Nerves, Anxiety, Heavy Air, Good Luck, Fright of Death/ Susto de la muerte, Spiritual Flowering, Good Business, Protection, Good Luck, Good Health / Leaves and Stems, fresh / Oral / 20g in 1 liter of water boiled 2 minutes and combined with 10g each of Pimpinela, Cedrón, Mejorana, Siempre Viva, Flores de Diamelas, Toronjil, Romero, Claveles and Orange Flowers. One liter a day or 3-4 cups a day after meals. **3.** Good Business, Protection, Good Luck, Good Health / Leaves and Stems, fresh / Seguro / Standard Seguro mixture.

Tillandsia multiflora

Bursera graveolens

Commiphora myrrha

Echinopsis pachanoi

Opuntia ficus-indica

Calceolaria rugulosa

BROMELIACEAE - *Tillandsia multiflora* Bentham var. *decipiens* (Andre) Sm.
Siempre Viva, Siempre Viva Roja
Herb, Andean, 1000-2500m
Use: 1. Depression, Heart, Nerves / Flowers, fresh / Oral / 10g in 1 liter of boiling water combined with 10g each of Toronjil, Clavelin and Lime Juice. Take 3 times a day for 1 month. **2.** Depression, Heart, Nerves / Whole plant, fresh / Topical / 20g per 5 liters of water boiled for 20 minutes. Bathe 3 times a week.

BURSERACEAE - *Bursera graveolens* (Kunth) Triana & Planchon
Palo Santo
Tree, Andean, 1000-3500m
Use: 1. Daño/Sorcery, Fright/Susto, Sorcery / Small Stems, Bark, Wood, dried / Topical / 3 tbsp per 3 liters of water combined with 10g each of Romero Blanco and Romero Castilla. Bathe 2-4 times a month. **2.** Cough, Flu, Bronchitis, Cold / Small Stems, Bark, Wood, dried / Oral / Boil 1 liter of water, then add 2 pieces of 5-10g of Palo Santo. Boil for 5 minutes. Cover and let sit for 3 minutes. Drink hot, 1 small glass 3 times a day for 2 days only. **3.** Dispelling negative energy from the house, Bad Shadow / Small Stems, Bark, Wood, dried / Incense / Big house: Use 250g of the herb. Small house: 20g combined with Romero Blanco, Romero de Castilla, Romero, Hierba de la Plata, Hierba de la Fortuna, Hierba de Oro, Incense (Copal), and Myrrh, every Tuesday and Friday as needed. For people: Patient must be naked with a piece of cloth tied to the neck. Place the jar with the smoking Palo de Santo under the person's feet letting the smoke rise. Also used in animal corrals mixed with Palo de Huaco to keep insects away.

BURSERACEAE - *Commiphora myrrha* (T. Nees) Engl.
Mirra (Myrrh)
Tree, resin, introduced
Use: Dispelling negative energy from the house / Latex, dried / Incense / Burn on charcoal and mix with 10g of Myrrh plus Palo Santo, Saumerio, and Romero. Burn as incense and spread smoke around the patient's house, 3 times a week: Tuesday, Friday, Tuesday. Repeat as necessary.

CACTACEAE - *Echinopsis pachanoi* (Britton & Rose) Friedrich & G. Rowley
San Pedro, Huachuma
Herb, Andean, Coastal, 0-3000m, cultivated
Use: 1. Ulcers, Hallucinogen, Enhancing Vision during rituals, Wounds caused by Daño/Sorcery, Bad Air/Mal Aire, Inflammation (general), Acne / Whole plant, fresh / Oral, Topical / Chop San Pedro in thin slices and boil in 4 liters of water from 12 noon to 6 PM. Cook on low fire and add water if necessary, 1 glass for the patient and 1 glass for the shaman per ritual session. Cannot eat fat, spices (such as aji), beans, fish or shellfish for 24 hours after drinking. Apply topically for wounds and acne. Patient should stay away from the sunlight for 24 hours. **2.** Washing hair, Fortifying hair / Whole plant, fresh / Topical / Rub pulp into the scalp.

CACTACEAE - *Opuntia ficus-indica* (L.) Miller
Tuna
Herb, Andean, Coastal, 0-3500m, introduced and cultivated
Use: 1. Diabetes / Fruits, fresh / Oral / Peel and extract Fruit. Drink the extract, 1 glass a day as needed. **2.** Hair Loss / Leaves, fresh / Topical / Cut a leaf in half. Boil each half in 3 liters of water for 20 minutes and drain. Wash hair with preparation and rub scalp as a shampoo. Bathe once a day for 4 days.

CALCEOLARIACEAE - *Calceolaria rugulosa* Edwin
Potito
Herb, Amazonian, Andean, Coastal, 0-4000m, weed
Use: Inflammation / Whole plant, fresh / Oral / Add 10g of plant material to 10g each of Verbena, Cola de Caballo, Pie de Perro, Amor Seco, Llantén, and 1 liter of water. Boil the mixture for 3 minutes. Drink warm. Take 1 cup, 3-4 times a day for 1 month.

Mammea americana

Centropogon argutus

Centropogon cornutus

Centropogon rufus

Lobelia decurrens

Siphocampylus angustiflorus

CALOPHYLLACEAE - *Mammea americana* L.
Mamey
Tree, Amazonian, Andean, Coastal, 0-1000m, cultivated
Use: 1. Diarrhea / Fruit peel, fresh / Oral / Boil 1 cup of water, then add 1/4 of the Mamey Fruit Peel. Cover and let sit for 2-3 minutes. Patient should drink warm solution, 1 glass 2-3 times a day for 2 days. **2.** Weight loss / Leaves, fresh / Oral / Boil 1 liter of water with 4 Mamey Leaves for 3-4 minutes. Drink 1 glass 3-4 times a day for about 1 month.

CAMPANULACEAE - *Centropogon argutus* E. Wimmer
Conchalay, Conchalalay
Shrub, Andean, 2000-3000m
Use: Fright/Susto, Air/Aire / Stems and Leaves, fresh or dried / Topical / 20g per 5 liters of water, boiled for 20 minutes. Bath: 1-3 times a month.

CAMPANULACEAE - *Centropogon cornutus* (L.) Druce
Raínga
Herb or Shrub, Amazonian, Andean, 0-1500m
Use: Bad Air/Mal Aire, Dissolve/remove tumors / Leaves and Stems, dried / Oral / Boil 100g of the plant in 1 cup of water. Drink cold, once a day. Must be followed by other treatments with other herbs.

CAMPANULACEAE - *Centropogon rufus* Wimm
Trinoso
Herb, Andean, 2000-3000m
Use: Intestine, Liver, Gallbladder, Tumors, Urinary Tract, Skin / Leaves and Stems, fresh or dried / Oral / Combine 10g of each of the following: Cadillo, Amor Seco, and Lampazo in 1/2 liter of water and boil for 5 minutes. Drink lukewarm, 1/2 cup 3 times a day for 20 days or as needed.

CAMPANULACEAE - *Lobelia decurrens* Cavaniles
Contolla
Herb, Andean, 1000-3500m
Use: Curing drug addicts. Causes vomiting and diarrhea. / Whole plant, fresh / Oral / Boil 1 liter of water, then add 5g of Contolla. Drink 1 cup a week for up to 1 month. Alternatively, empty a cigarette 75%, fill with 50% ground Contolla and refill remaining 25% with tobacco. Smoke.

CAMPANULACEAE - *Siphocampylus angustiflorus* Schlechtendal
Contoya, Hierba de Envidia, Contolla
Vine, Andean, 1500-3500m
Use: 1. Purgative / Flowers, Leaves and Stems, fresh / Oral / Boil 5g in 1/2 cup of water. Drink once a month. **2.** Daño/Sorcery, Cast away Envy / Flowers, Leaves, and Stems, fresh / Topical / Boil 10g of Contolla in 10 liters of water for 20 min mixed with other herbs as specified by the *curandero* for Good Luck. Bathe 3 times a week.

Siphocampylus cutervensis

Siphocampylus tupaeiformis

Celtis pubescens

Capparis crotonoides

Capparis scabrida

Dipsacus jallonum

CAMPANULACEAE - *Siphocampylus cutervensis* A. Zahlbr.
Conchalay Blanco, Conchalalay Blanco
Herb, Andean, 2000-3000m
Use: 1. Concussions / Leaves, fresh / Topical / Plaster: once a month. **2.** Fright/Susto, Daño/Sorcery / Leaves, dried / Topical / One handful in 3 liters of boiled water combined with Conchalay Colorado, Huaminga, Vinegar and 7 Espiritus. Two Baths a week in agreement with what the Mesa indicates. Limpia: 2 times a month.

CAMPANULACEAE - *Siphocampylus tupaeiformis* Zahlbr.
Cochaya
Herb, Andean, 3000-4000m, weed
Use: Guarding the house and land. / Whole Fruit, fresh / Amulet / Do not ever cut the plant. Use it in its natural form by planting it in the area around your house. Always plant Cochaya close to a San Pedro. Always have a few on your property. If a thief comes onto your property, this plant will start to release snakes all around and tie the person up without ropes. A thief can go crazy. This plant always detects who is doing bad things and recognizes its owner and its family.

CANNABACEAE - *Celtis pubescens* (Humb. & Bonpl.) Spreng.
Palo Huaco, Palo Blanco
Tree, Amazonian, Andean, 500-1500m, weed
Use: Fertility, Sexual Potency, Arthritis, Bronchitis, Muscle Pain, Blood Circulation, Hemorrhages (healing) / Bark, dried / Oral / Mix Palo Huaco, Palo Sangre, Chuchuhuasi, Huanaco, Huevo del Angelote, Pacra, Pollen, Miel de Palo, Honey, Cascarilla, and Huanarpo Macho in 1 bottle of Wine or Tequila. Let mixture sit for 1 week. Drink cold, 1 small wine glass 3 times a day until bottle is finished. Patient can repeat the treatment.

CAPPARIDACEAE - *Capparis crotonoides* (Kunth) Iltis & Cornejo
Cimuro, Simuro, Bichayo
Tree, Andean, Coastal, 0-1000m
Use: 1. Bronchitis / Flowers, fresh / Oral / Boil 10 Flower buds in 1/2 cup of water for 2 minutes. Patient should drink warm solution and stay inside the house during treatment. Drink 1 cup a day for 8 days. **2.** Arthritis, Rheumatism / Leaves, fresh / Topical / Boil 2kg of Bichayo Leaves in 5 liters of water for 30 minutes. The patient must be naked with a towel over his head in a closed room. Patient must make deep inhalations. Bath should last about 1/2 hour. Bathe every 6 days, 2 times only. **3.** Cold, General pain: muscular, bone, etc. / Leaves, fresh / Topical / Crush 20 Leaves of Bichayo and place crushed Leaves on affected area. Massage the area with Leaves. Patient should not go out during treatment. **4.** Bad Air/Mal Aire, Colds / Leaves, fresh / Topical / Add 20g of plant material to 4-5 liters of water. Boil the mixture for 5-6 minutes. Do not ingest the mixture. Bathe 2-3 times as needed.

CAPPARIDACEAE - *Capparis scabrida* Kunth
Zapote
Tree, Andean, Coastal, 0-2500m
Use: Inflammation (general), Heart Palpitations, Liver, Anxiety Reduction, Increased milk production in cows / Fruit, fresh / Oral / Blend Fruit and collect extract. Drink unheated, 1 glass a day for 4 days. Also used as glue extracted from the trunk of the tree.

CAPRIFOLIACEAE - *Dipsacus jallonum* L.
Cardo Santo
Herb, Andean, Coastal, 0-3000m, weed, introduced
Use: Diabetes, Liver, Cholesterol / Whole plant, fresh / Oral / 3-5g in 1 liter of boiling water mixed with herbs that are used for the same conditions. Drink 3 times a day.

Lonicera japonica

Scabiosa atropurpurea

Valeriana microphylla

Valeriana niphobia

Valeriana plantaginea

Valeriana rigida

CAPRIFOLIAEAE - *Lonicera japonica* Thunberg
Madre Selva
Shrub, Andean, 2000-3000m, introduced
Use: Depression, Heart, Lovesickness, Nerves, Epilepsy, Psychological Suffering / Whole plant, fresh or dried / Oral / Boil 1 liter of water, then add 10g of Madre Selva. Drink 3-4 cups a day for 1-3 months or as needed. Epilepsy is characterized by heart pains, frequent falling to the ground and loss of consciousness.

CAPRIFOLIACEAE - *Scabiosa atropurpurea* L.
Ambarina, Ambarina Negra, Flor de Ambarina, Ambarindas
Herb, Andean, Coastal, 0-3000m, weed, introduced
Use: 1. Whooping cough, Cold, Cough, Bronchitis, Blood Cleansing, Compulsive Cough / Flowers, fresh / Oral / Boil 1 liter of water with 20g of the plant material and Veronica, Hierba del Toro, Moradilla, Lancetilla, and Hierba de la Rabia. Drink hot 3 times a day as long as the ailment lasts. **2.** Menstrual regulation / Flowers, fresh / Inhaled / Chop and mix with maternal milk. Inhale 5g daily for 8 days through the nose and also take orally.

CAPRIFOLIACEAE - *Valeriana microphylla* Kunth
Hierba de la Fortuna
Herb, Andean, 3000-400m
Use: Fragrance, Good Luck / Whole plant, fresh or dried / Topical / 10g in 1 liter of water, 2 baths a month in the evening.

CAPRIFOLIACEAE - *Valeriana niphobia* Briquet
Botón de Oro
Herb, Andean, 3500-4500m
Use: 1. Good Luck / Whole plant, fresh or dried / Topical / Boil 3 liters of water with 10g of Botón de Oro and 10g each of Hierba de la Justicia, Hierba del Halago, Hierba de la Plata, Hierba de la Fortuna, Dolar and Sigueme Sigueme for 3 minutes. Add Agua Florida, Tabú, White Sugar and Lime Juice. Bathe 3 times a week on Tuesday, Friday and Tuesday. **2.** Good Luck / Whole plant, fresh or dried / Seguro / Prepare with perfumes and the typical Seguro herbs. Refill with perfumes as needed. Keeps its power as long as it is kept full.

CAPRIFOLIACEAE - *Valeriana plantaginea* Kunth
Hórnamo Morado, Hórnamo Caballo
Herb, Andean, 3500-4500m
Use: 1. Bad Air/Mal Aire, Purgative, Laxative / Leaves and Stems, fresh / Oral / Boil 10g of Hórnamo Morado with 1/2 cup of water for 2 minutes. Patient should drink cold solution, 1/2 cup once only. **2.** Protection / Leaves and Stems, fresh / Topical / Boil 3 liters of water for 10 minutes with 100g of Hórnamo Morado and 10g each of: Misha Blanca, Misha Colambo, Misha Galga, Misha Morada, Misha Roja, Misha Rosada, and Toro Maique. Recite a prayer. Patient should rub self with herbs. When the bath is finished, the patient should not rinse or use a towel, but air dry only

CAPRIFOLIACEAE - *Valeriana rigida* Ruiz. & Pav.
Hórnamo Estrella, Siete Sábios, Valeriana Estrella, Valeriana, Hierba de la Estrella
Herb, Andean, 2500-4500m
Use: 1. Fragrance, Good Luck / Stems, fresh / Seguro / Mix with other herbs of strength, herbs of luck for Seguro. **2.** Spiritual Flowering, Good Luck, Bad Air/Mal Aire, Success / Stems, fresh / Topical / Boil 20g per 5 liters of water for 20 minutes. Mix with other herbs of strength and herbs of luck. Bathe 3 times a week. **3.** Insomnia, Relaxant, Sleeplessness, Nerves, Headache, Menopause / Stems, fresh / Oral / Boil 1 liter of water, then add 10g of Valeriana Estrella. Drink 4 times a day as needed. Children cannot take it often; can only begin at 6 years of age. **4.** Contusions, Mental Disorders, Schizophrenia, Cerebral Pain / Stems, fresh / Topical / Combine with Timolina, Vinegar, Agua Florida and Árnica. Put on the back of the head or afflicted area. Leave on at night. **5.** Spiritual Flowering, Good Luck, Bad Air/Mal Aire, Success / Stems, fresh / Topical / In a bottle place 1g of each of the following: Hierba del Lucero, Hierba Estrella, Ambrocilla, Señorita, Caballero, Pega Pega, Siempre Viva, Carpintero, Waime Waime, Piri Piri (Hembra y Macho), Hierba del Buen Querer, Hierba del Oro, Hierba de la Plata, Hierba del Halago, Sigueme Sigueme, and Hierba del Negocio. Add some

Carica papaya

Jacartia digitata

Dianthus caryophyllus

Dianthus caryophyllus

Stellaria media

Salacia multiflora

drops of the following perfumes: Cariño, Dios de la Huaringa, Dios de la Felicidad, San Antonio, Macumba Pusanga, Gran Jefe, Mil Flores, Llama Plata and Ekeko. Recite a prayer invoking the name of the patient, owner of the Seguro. Spray and rub the patient with the mixture for Good Luck, Tuesdays and Fridays. **6.** Spiritual Flowering, Good Luck, Bad Air/Mal Aire, Success / Stems, fresh / Topical / Mix a total of 50g of Hierba del Lucero, Hierba de la Estrella, Ambrocilla, Señorita, Caballero, Pega Pega, Siempre Viva, Carpintero, Waime Waime, Piri Piri (Hembra y Macho), Hierba del Buen Querer, Hierba del Oro, Hierba de la Plata, Hierba del Halago, Sigueme Sigueme, and Hierba del Negocio. Boil in 5-7 liters of water for 20 minutes, then add a bit of the following perfumes: Cariño, Dios de la Huaringa, Dios de la Felicidad, San Antonio, Macumba Pusanga, Gran Jefe, Mil Flores, Llama Plata, and Ekeko and let it cool before bathing. Bathe 2 times (Tuesdays and Fridays only) every 3 months.

CARICACEAE - *Carica papaya* L.
Papaya
Tree, Amazonian, Andean, 0-3000m, cultivated
Use: 1. Stomach Parasites, Laxative, Anti-venom, Reverse effects of Poison / Seeds and Fruit Peel, fresh / Oral / Blend 10 liters of water, 1 cup of Seeds and a whole peel of a small Papaya. Drink while fasting, 1 glass 3 times a month. This will make you vomit and defecate a great deal. **2.** Inflammation of the liver / Fruit / Oral / 1 small bowl every day for 1 week before breakfast. Fast.

CARICACEAE - *Jacartia digitata* (Poepp. & Endl.) Solms-Lang.
Contra Hechizo
Tree, Amazonian, Andean, 0-3000m
Use: 1. Purgative (Daño/Sorcery), Laxative for people who suffer from colic pain and gases / Root, fresh / Oral / Boil a 25cm portion of the Root in 3 liters of water for 20 minutes or crush 200g of Contra Hechizo and add 50g of sugar. Drain the extract. Drink 1 glass during a curing session. **2.** Acne / Root, fresh / Topical / Crush Stems of Contra Hechizo and drain extract. Apply on face or affected areas as a cream twice a day: AM and before going to bed for 6 days or until Acne disappears.

CARYOPHYLLACEAE - *Dianthus caryophyllus* L.
Clavel, Clavelina, Clavel de la Costa
Herb, Andean, 2000-4000m, introduced and cultivated
Use: 1. Lovesickness/Mal de amor, Sentimentality, Heart, Nerves, Good Luck, Insomnia / Whole plant, fresh / Topical / Boil 7 entire plants with Hierba de la Plata, Hierba de la Justicia, Ruda and Romero in 3 liters of water for 5 minutes. Bathe as needed on special days depending on the phases of the moon, 1-3 times a month. **2.** Lovesickness/Mal de amor, Sentimentality, Heart, Nerves, Good Luck, Insomnia / Whole plant, fresh / Oral / Drink 50g of Claveles (white, red, yellow and purple petals) with 5g of sugar and 1/2 cup of water boiled for 2 minutes, 3-4 cups a day for 1 month.

CARYOPHYLLACEAE - *Dianthus caryophyllus* L.
Clavel Serrano
Herb, Andean, 2000-4000m, introduced and cultivated
Use: Heart disease, Heart Palpitations / Whole plant, fresh / Oral / Combine 50g of the plant material, 50g of Huamanripa and 1 cup of water. Boil the mixture for 5 minutes. Drink the mixture cold. Take 1/4 cup once a day for 30 days.

CARYOPHYLLACEAE - *Stellaria media* (L.) Criollo
Tripa de Cuy
Herb, Andean, Coastal, 0-4000m, weed
Use: Inflammation of the Kidneys, Renal Disease / Whole plant, fresh / Oral / Add 10g of plant material to 1 liter of water, Malva, Amor Seco, Chacur and Unguia. Drink 1 cup 3-4 times a day for 1 month.

CELASTRACEAE - *Salacia multiflora* (Lam.) DC.
Bejuco de Montaña
Liana, Amazonian, Andean, 0-1000m

Hedyosmum racemosum

Couepia guianensis

Clethra castaneifolia

Clusia minor

Cuscuta foetida

Ipomoea batatas

Use: Nervous system / Seeds and Stems, fresh or dried / Topical / Boil 200g per 3 liters of water. Take 2 baths a month.

CHLORANTHACEAE - *Hedyosmum racemosum* (R. & P.) G. Don.
Masamoche, Asancito, Asarcito, Asarquiro, Choleta
Tree, Amazonian, Andean, 0-2500m
Use: Bronchitis, Cold, Cough, Asthma, Rheumatism, Bone Pain, Nervous system / Bark, dried / Oral / Use outside of Bark, 8-10g per 2 liters of water boiled for 20 minutes. Drink as needed. Alternatively, use 30g per 2 bottles of alcohol mixed with Chuchuhuasi, Cascarilla, 7 Raices and Huayacanes, then allow to sit for 8 days. Drink as needed.

CHRYSOBALANACEAE - *Couepia guianensis* Aubl.
Acharachango, Charachango, Ashango
Tree, Amazonian, 0-500m
Use: **1.** Fright/Susto, Mal Aire/Bad Air, Nervous System, Enchantment, Sorcery, Getting rid of Daño/Sorcery, Epilepsia / Seeds, dried or fresh / Topical / Bath: 20 seeds per 5 liters of water ground and boiled, 3 times a week or twice a month, Tuesday and Friday. **2.** Fright/Susto, Mal Aire/Bad Air, Nervous System, Enchantment, Sorcery, Getting rid of Daño/Sorcery, Epilepsia / Seeds, dried or fresh / Oral / Beverage: 7-15 seeds in 1 liter of water crushed and macerated for 8 days. Drink 3-4 small cups a day for 7 days. Seeds can be macerated in alcohol for 5 days, then drunk, 5g 3 times a day.

CLETHRACEAE - *Clethra castaneifolia* Meisn.
Hierba del Olvido
Tree, Andean, 1000-2500m
Use: Dominating judgment (ritual), Dominating legal problems (ritual) / Leaves and Stems, dried / Seguro / Put together in a cloth 10g of Hierba del Olvido, 10g of Hierba del Demonio and 10 g of Hierba de la Justicia. Seal and pray. Patient must carry the bag and pray.

CLUSIACEAE - *Clusia minor* L.
Chusgón
Tree, Amazonian, 0-500m
Use: Nerves, Heart disease / Whole Fruit, fresh or dried / Oral / Boil 1 liter of water, then add 10g of Chusgón, Pimpinela, Manzanilla, Toronjil, Barrojo, Romero, and Chancas de Comida. Let it sit for 2-3 minutes. Patient should drink warm solution, 1 cup 4 times a day as needed.

CONVOLVULACEAE - *Cuscuta foetida* H.B.K.
Yodo
Parasitic herb, Andean, 2500-4000m
Use: **1.** Goiter / Whole plant, fresh / Oral / 5g of boiled in 1 cup of water. Drink 1 cup a day. **2.** Goiter / Whole plant, fresh / Topical / 5g boiled in 1 cup of water. Apply as Poultice.

CONVOLVULACEAE - *Ipomoea batatas* (L.) Lamarck
Camote (Sweet potato)
Herb, Amazonian, Andean, 0-2500m, introduced and cultivated
Use: Promoting lactation in women after giving birth / Whole plant, fresh / Oral / First wipe the patient with the plant, making the sign of the cross over the chest. Then wash and place plant in 1 cup of water and heat. Drink warm 1 cup 2 times a day for 2 days.

Ipomoea pauciflora

Echeveria peruviana

Citrullus lanatus

Cucumis dipsaceus

Cucumis sativus

Cucurbita maxima

CONVOLVULACEAE - *Ipomoea pauciflora* M. Martens & Galeotti
Huanarpo
Shrub, Amazonian, Andean, 0-3000m
Use: Chills, Colds / Whole plant, fresh / Oral / Put together in a bottle of Cañazo or Yonque (Sugarcane Alcohol) 20g of the plant material plus 20g of Cascarilla, Diego Lope and Hualtaco. Let it sit for 8 days. Drink cold 1 small cup once a day or as needed (maximum 2 days only).

CRASSULACEAE - *Echeveria peruviana* Meyen
Pinpin, Siempre Viva, Rosa Berta, Haya Rosa
Herb, Andean, 2000-3500m
Use: 1. Kidney Inflammation, Nerves, Heart, Liver Inflammation / Leaves, fresh / Oral / 2 Leaves per 1/2 liter of water boiled for 5 minutes. Drink 1 liter daily. **2.** Ear Ache / Leaves, fresh / Topical / Heat over fire then squeeze Leaf for liquid. Place 10 drops in ear everyday or as needed.

CUCURBITACEAE - *Citrullus lanatus* (Thunberg) Matsumura & Nakai
Sandía (Watermelon)
Vine, Amazonian, Andean, Coastal, 0-3000m, introduced and cultivated
Use: Blood Circulation, Refresh Heart / Bark, dried / Oral / Peel the Fruit. Take the pulp and blend. Drink cold, 1 glass a day while fasting for 20 days.

CUCURBITACEAE - *Cucumis dipsaceus* Ehrenb.
Vine, Amazonian, Andean, Coastal, 0-3000m, introduced and cultivated
Jaboncillo del Campo, Jaboncillo, Patito de Campo
Use: Dandruff, Adding shine and beauty to hair, Hair Loss (prevention), Stopping baby from breastfeeding / Fruit, fresh / Topical / Use Fruit as a shampoo and soap. Open the Fruit and rub the scalp with the seeds. Bath: Use the Fruit to rub the whole body. Rub 1/2 of the Fruit on the breast of a breastfeeding mother. Wash 2-3 times a week as needed.

CUCURBITACEAE - *Cucumis sativus* L.
Pepinillo
Vine, Amazonian, Andean, Coastal, 0-3500m, weed, cultivated
Use: Burn Fat, Weight Loss, Liver Inflammation, Indigestion, Heartburn, Intestinal Acidity / Whole Fruit, fresh / Oral / Remove the peel, finely chop the Fruit and add a few drops of olive oil and a pinch of salt. Extract juice or eat as salad. Drink cool while fasting, 1 glass a day as needed or eat a small serving. Oral: once a day for a week.

CUCURBITACEAE - *Cucurbita maxima* Duchesne
Zapallo
Vine, Amazonian, Andean, Coastal, 0-3000m, cultivated
Use: 1. Preventing Miscarriage, Inflammation (general), Anxiety / Flowers and joints of Stems, fresh or dried / Oral / Boil 10 Flowers with Leaves and Stems in 1/2 liter of water for 15 minutes. Drink 1 cup 2 times a day or as much as wanted for 2 days. **2.** Heart Disease / Seeds, dried / Oral / Oral only in the morning with a glass of water. Take 10 seeds the first day. Reduce amount by one seed each day until, on the 10th day, you take only 1. Take a break of one week before repeating process. Process may last 2 weeks to a month. **3.** Warts, Moles / Leaves, fresh / Topical / Chop 3-4 big Leaves. Place chopped Leaves on affected area and cover with a piece of plastic, then a piece of cloth. Apply for 2 days, then remove.

Cucurbita moschata

Cyclanthera pedata

Sechium edule

Sicana odorifera

Sycos baderoa

Cupressus lusitanica

CUCURBITACEAE - *Cucurbita moschata* Duch.
Zapallo
Vine, Amazonian, Andean, Coastal, 0-3000m, cultivated
Use: 1. Preventing Miscarriage, Inflammation (general), Anxiety / Flowers and joints of Stems, fresh or dried / Oral / Boil 10 Flowers with Leaves and Stems in 1/2 liter of water for 15 minutes. Drink 1 cup 2 times a day or as much as wanted for 2 days. **2.** Heart Disease / Seeds, dried / Oral / Oral only in the morning with a glass of water. Take 10 seeds the first day. Reduce amount by one seed each day until, on the 10th day, you take only 1. Take a break of one week before repeating process. Process may last 2 weeks to a month. **3.** Warts, Moles / Leaves, fresh / Topical / Chop 3-4 big Leaves. Place chopped Leaves on affected area and cover with a piece of plastic, then a piece of cloth. Apply for 2 days, then remove.

CUCURBITACEAE - *Cyclanthera pedata* (L.) Schrad.
Caihua
Vine, Amazonian, Andean, Coastal, 0-3000m, cultivated
Use: Brain Tonic, Renal Inflammation, Brain Disease, Headache / Whole plant, fresh / Oral / Blend/ liquefy 3 Caihuas, 10g Rutabaga, 10g of Carrots and 10g of Radishes; 1 8oz glass a day for 15 days in the morning while fasting. Drink cold.

CUCURBITACEAE - *Sechium edule* Swartz.
Caihua Chilena, Mochera, Caihua
Vine, Amazonian, Andean, Coastal, 0-3000m, introduced and cultivated
Use: Blood Circulation, Varicose Veins (Blood Clot Dilution) / Fruits, fresh / Oral / 1/2 of the Fruit chopped with 10g of Moradilla. Boil in 1 liter of water for 3-5 minutes. Drink 3-4 times a day for 1 month.

CUCURBITACEAE - *Sicana odorifera* (Vell.) Naud.
Secana
Vine, Amazonian, Andean, Coastal, 0-3000m, introduced and cultivated
Use: Jealousy / Whole plant, fresh / Amulet / Sleep with plant material and wash it every morning. Do not eat the plant. At the end of the treatment, the Secana will be wrinkled. Dispose of it in a distant place. Keep the Secana for 7 days and 7 nights.

CUCURBITACEAE - *Sycos baderoa* H. et A.
Fuque
Vine, Amazonian, Andean, 0-3000m, weed
Use: 1. Snake bites / Seeds, dried / Oral / Grind 100g of Fuque and add 100g of Cuncuno and 5g of cooking oil. Drain with a bit of water. Drink cold at 6AM in the morning while fasting, 1/2 a small glass for 4 days. **2.** Getting rid of skin Mold / Leaves, fresh / Topical / Boil 100g of the Leaf with 1 liter of water until all the liquid is gone. Place affected area over the boiling pot and let the steam touch it. Then take the Leaves and place them on top of the affected area while hot, once a day for 15 days. Use steam therapy for 20 min and leave the Leaf on top of the mold for 3 hours.

CUPRESSACEAE - *Cupressus lusitanica* Miller
Cipre, Cipres
Tree, Andean, 2500-4000m, introduced and cultivated
Use: 1. Vaginal Hemorrhage, Hemorrhage / Whole plant, fresh / Oral / Boil 3 Branches in 1 liter of water for 20 minutes. Drink 2 times a day for 1 week. **2.** Acne, Hair Loss / Whole plant, fresh / Topical / Boil 1/2 liter of water containing 10g of Cipres for 3 minutes. Bathe and wash face with warm solution. Do not rinse after the solution is put on the face. Apply 3 times a day for 1 week. For Hair Loss boil 1 liter of water containing 10g total of Amor Seco, Romero, Cola de Caballo and Cipres for 3 minutes. Wash hair with cold solution once a day as needed.

Cyperus articulatus

Kyllingia pumila

Oreobolos obtusangulus

Scirpus californicus

Dioscorea tambillensis

Dioscorea trifida

CYPERACEAE - *Cyperus articulatus* L.
Bastón de San José, Varita de San José, Pura Pura
Herb, Amazonian, Andean, Coastal, 500-2500m
Use: 1. Protection / Whole plant, fresh / Seguro / Add plant material to Hierba de la Plata, Hierba de la Justicia, El Dolar, Carpintero, Chupa Flor, Señorita, Hierba de Oro, Fortuna, Agua Florida, Ramillete de Novia, Lime Juice, White Sugar and Holy water. Put all materials in a bottle: first the herbs, then the remaining materials. Cover the mixture. Only the patient can touch the Seguro. Keep in the house. **2.** Spiritual Flowering / Whole plant, fresh / Topical / Boil the plant material with Hierba de la Plata, Hierba de la Justicia, Dolar, Carpintero, Chupa Flor, Señorita, Hierba del Oro, Fortuna, Roses (various kinds), Ruda (Hembra and Macho) and 3 liters of water. Boil the mixture for 10 minutes to concentrate the solution. The patient is rubbed with Flowers while bathing in the lukewarm mixture. Bathe Tuesday, Friday and the following Tuesday (1 cycle). The bath should be just before noon, at 3PM or at midnight. Patients should bathe for 3 cycles.

CYPERACEAE - *Kyllingia pumila* Michx.
Hierba de la Golondrina
Herb, Amazonian, Andean, 0-3000m, weed
Use: Heart, Nerves, Luck, Daño/Sorcery against a woman / Whole plant, fresh / Topical / Boil 30g in 6 liters of water for 20 minutes with other Good Luck Herbs. Bathe 3 times a week.

CYPERACEAE - *Oreobolos obtusangulus* T. Koyama
Hierba Chupaflor, Hierba de la Suerte, Hierba del Carpintero
Herb, Andean, 3000-4000m
Use: 1. Success, Bad Air/Mal Aire, Good Luck, Work, Aphrodisiac, Good Business, Protection, Good Fortune, Good Health / Leaves, dried / Topical / Boil 20-50g in 3-5 liters of water for 20 minutes. Mix with Good Luck Herbs and Herbs of Strength. Bathe 3 times a week. Alternative mixture for Spiritual Flowering, see below, once only.
2. Success, Bad Air/Mal Aire, Good Luck, Work, Aphrodisiac, Good Business, Protection, Good Fortune, Good Health / Leaves, dried / Seguro / Standard Seguro mixture.

CYPERACEAE - *Scirpus californicus* (C.A. Meyer) Steudel subsp. *tatora* (Kunth) T. Koyama
Balsa, Totora
Herb, Andean, Coastal, 0-3500m, cultivated
Use: 1. Athlete's foot / Whole plant, dried / Topical / Burn whole and create ash. Apply powder from burned ashes to infected area and scrub 3 times a day as needed. **2.** Hangover, Evil Eye/Mal Ojo / Heart of the Stem, fresh / Amulet / Make a small brush out of the insides of the Stems. Use the brush to make the sign of the cross on the patient followed by rubbing. After use, burn the plant. If it falls apart, it means the child has Evil Eye/Mal Ojo. This is used only for children. **3.** Fever, Colds / Whole plant, dried / Oral / To 1/2 cup of water add 10g of Totora and boil for 3 minutes. Drink cold, 1/2 cup a day for 8 days.

DIOSCOREACEAE - *Dioscorea tambillensis* Kunth
Papa Semitona
Herb, Andean, 1500-2500m, cultivated
Use: Kidney Inflammation, Ovary Inflammation, Liver Inflammation / Tuber, fresh / Oral / Boil 1 liter of water and add 1 Papa Semitona (10g) along with 20g total of Amor Seco, Chacur, Cola de Caballo, Pie de Perro, Verbena, Linaza, and toasted Cebada. Boil for 2-3 min. Drink lukewarm. One can also add Sugar or Limes for flavor. Drink 1 cup 3 times a day for 1 month.

DIOSCOREACEAE - *Dioscorea trifida* L.f.
Papa Madre, Papa Pacta
Herb, Amazonian, Andean, 0-1000m
Use: 1. Inflammation, Uterine Disease and Discharge, Cysts, Internal Inflammation, Uterine Cancer, Ovary Inflammation, Vaginal Discharge, Kidney Inflammation / Tuber, fresh / Oral / Boil 1 liter of water with 1/4 of a big tuber for 5 minutes along with Flor Blanca, Purenrosa, Pacharosa and 10g of watercress. Drink lukewarm 3-4 times a day for 1 week or as needed, especially for children 3 months - 5 years. **2.** Fungus, Vaginal Cleansing, Uterine Cancer, Wound Cleansing / Tuber, fresh / Topical / Boil 20g in 1 liter of water for 20 minutes. Combine with Matico, Malva, and Tara. Drink 1 liter a day for 15 days.

Vallea stipularis

Ephedra americana

Equisetum bogotense

Equisetum giganteum

Bejaria aestuans

Gaultheria erecta

ELAEOCARPACEAE - *Vallea stipularis* L.f.
Chuingue
Tree, Andean, 1500-4500m
Use: Daño/Sorcery, Hearing problems, Deafness / Leaves, fresh or dried / Topical / Use 1 handful in 3 liters of water mixed with Timolina for Limpia. Alternatively, mix maternal milk together with the Leaf and put in the ears.

EPHEDRACEAE - *Ephedra americana* Humb. & Bonpl. ex Willd.
Diego Lopez, Suelda con Suelda
Shrub, Andean, 500-4500m
Use: 1. Bruises, Fractures, Broken Bones, External Injuries / Whole plant, fresh or dried / Oral. Boil 5g in 2 liters of water for 5 minutes. Drink 3 times a day, 1-2 cups a day for no more than 2 weeks. **2.** Bruises, Fractures, Broken Bones, External Injuries / Whole plant, fresh or dried / Topical / Grind plant and mix with the fat of a male donkey and place around limb. Wear every day for the entire day until bone mends.

EQUISETACEAE - *Equisetum bogotense* H.B.K. (Kunth)
Cola de Caballo
Herb, Amazonian, Andean, Coastal, 0-4500m, weed
Use: 1. Kidney Inflammation, Wounds (Cleansing), Stomach, Urinary Tract, Kidneys, Kidney Stones, Inflammation (general) / Stems, dried / Oral / Boil 5g in 1 liter of boiling water mixed with 10g each of Overo, Lancetilla, Chante, Achiote and Zarzaparrilla. Drink 1 liter daily for 1 month. **2.** Kidney Inflammation, Wounds (Cleansing), Stomach, Urinary Tract, Kidneys, Kidney Stones, Inflammation (general) / Stems, dried / Topical / Wash Wounds with 10g total of this herb mixed with Pie de Perro, Chacur, and Untia boiled in 1 liter of water.

EQUISETACEAE - *Equisetum giganteum* (Wedd.) Ulbrich
Limpia Plata, Cola de Caballo, Tembladera
Herb, Andean, 500-2500m
Use: 1. Arthritis, Kidneys, Hemorrhages, Menstrual inflammation, Internal and External Inflammation, Prostate, Kidney stones / Whole plant, fresh / Oral / Boil 10-20g in 1 liter of water combined with Verbena, Manzanilla, Chacur, Unquia, Espiga de Maiz, Paja Blanca, Berro, Pata de Perro, Papa Madre, Pelo de Choclo and other herbs that are good for the Kidneys. Mix with Chante and Ochote for Prostate. Boil 2-5 minutes and drink 4 cups a day for 1 month. **2.** Wounds (cleansing) / Whole plant, fresh / Topical / Boil 10g of Limpia Plata with 1 liter of water. Combine with 10g each of Chacur, Verbena, Hierba Santa, and Llantén. Clean wound with Leaves. Clean excess with gauze. Use lukewarm water twice a week until the Wound heals. Plant should be used to wash wounds, not as poultice.

ERICACEAE - *Bejaria aestuans* Mutis ex L.
Cadillo, Payama, Hierba de la Postema, Purenrosa, Hierba de la Postema Rosada, Hierba del Buen Querer
Shrub or Tree, Andean, 500-3000m
Use: 1. Diabetes, Liver, Prostate, Allergies, Menstrual Regulation, Blood related illnesses, Kidney Inflammation, Uterine Inflammation, Liver Inflammation, Inflammation of the Bladder, Cysts, Ovarian Inflammation, Blood Pressure, Menstrual Pain, Inflammation (general), Spiritual Flowering / Flowers, Leaves and Stems, fresh or dried / Oral / Boil 5g in 1 liter of water for 5 minutes with Flor Blanca, Papa Madre, Flor de Arena, Guayusa, Pasuchaca, Malva, Amor Seco, Verbena, Llantén, Cola de Caballo, Chumbiaura, Palo de China, Huaminga, and Quinua Agrio. Drink 3 cups daily or 1 liter a day for 1-3 months. **2.** Good Luck, Good Business, Protection, Good Fortune, Good Health / Flowers, Leaves and Stems, fresh or dried / Seguro / Standard Seguro mixture.

ERICACEAE - *Gaultheria erecta* Vent.
Mullaca Mistura, Mullaca, Mullaca Real
Shrub, Andean, 500-4000m
Use: Bronchitis, Asthma / Whole plant, fresh or dried / Oral / Boil 1 liter of water and add 10g of Mullaca. Include 10g of each of the following: Humanaripa, Escorcionera, Eucalyptus, Matico, and Veronica. Drink 1 cup 3 times a day for 1 month.

Gaultheria reticulata

Paepalanthus ensifolius

Erythroxylon coca

Escallonia pendula

Acalypha villosa

Chamaesyce hypericifolia

ERICACEAE - *Gaultheria reticulata* Kunth
Toromaique, Maique, Maique Candela, Toromaique Amarillo, Toromaique Verde, Gavilán Maique Amarillo, Gavilán Maique Verde
Shrub, Andean, 1000-4500m
Use: 1. Hallucinogen, Cold, Arthritis, Bronchitis, Rheumatic Pain, Bone Pain, Bad Air/Mal Aire, Wounds, Ulcers, Sores. / Whole plant, fresh / Topical / Boil 50g for 20-30 minutes in 7 liters of water and mix with other Maiques (7 varieties), i.e., 10g each of: Misha Blanca, Misha Colambo, Misha Galga, Misha Morada, Misha Roja, Misha Rosada, and Toromaique. Recite a prayer. Bathe 3 times a week. Bathe the patient in the mixture while rubbing him/her with the herbs. Afterwards, rinse the patient in water, and allow him/her to air dry. **2.** Purgative / Leaves, dried / Oral / Boil 3g of ground Leaves in 1 cup of water and take once a month in the morning before breakfast. **3.** Protecting the house, Protecting the patient / Whole plant, fresh / Seguro / Add 10g of plant material to Hierba de la Plata, Hierba de la Fortuna, Hierba del Oro, Carpintero, Chupa Flor, Señorita, Condores, Trenzadilla, Agua Florida, Tabu Perfume, Lime Juice, Agua Bendita, and Sugar. Place all ingredients into 1 bottle. **4.** Protecting the house, Protecting the patient, Rituals / Whole plant, fresh / Incense / Smoke the area using a counterclockwise motion as needed. **5.** Wounds, Cysts, Wounds from Sorcery / Whole plant, fresh / Topical / Boil 50g of plant material in 1/2 cup of water for 5 minutes. Apply to affected area once a day until it heals.

ERIOCAULACEAE - *Paepalanthus ensifolius* (Kunth) Kunth
Madriguera
Herb, Andean, 3000-4000m
Use: Make a business successful, To have control of employees / Whole plant, fresh / Seguro / 1/5 of plant per Seguro.

ERYTHROXYLACEAE - *Erythroxylon coca* Lam.
Coca
Shrub, Amazonian, Andean, 0-1500m, cultivated
Use: Cold, Cough, Inflammation of the throat, Induce child birth, Strength for woman during childbirth, Helping delivery of newborn, Alertness, Ritual practices / Leaves, dried / Oral / Add 5g of the Leaf to 1 cup of water. Boil the mixture for 3-4 minutes, then let it cool. Gargle 3 times a day for 2 days. Drink 1 cup before bed for 2-3 days. Alternatively, wash and chew about 5g of Leaves at a time.

ESCALLONIACEAE - *Escallonia pendula* (R. & P.) Pers.
Chuque
Tree, Andean, 2000-3000m
Use: 1., Arthritis, Bone Pain, Sorcery/Daño, Rheumatism, Fright/Susto of Death / Leaves, dried / Topical / Boil 1 bundle in 3 liters of water. Can combine with Huaminga, Chingue, Ishpinguillo, Ajenco, and 7 Espiritus. Bathe once a week. **2.** Arthritis, Bone Pain, Sorcery/Daño, Rheumatism, Fright/Susto of Death / Leaves, fresh / Topical / Poultice: Do not mix with other plants. Apply 3 times a week.

EUPHORBIACEAE - *Acalypha villosa* Jacq.
Chilca Dulce
Shrub, Andean, 500-2000m, weed
Use: Liver Inflammation, Blood Detoxification / Whole plant, fresh or dried / Oral / Put 10g of the plant in 1 liter of water and boil for 3-5 minutes. Drink 3 times a day for 2 weeks.

EUPHORBIACEAE - *Chamaesyce hypericifolia* (L.) Millspaugh
Lecherita, Lechera
Herb, Amazonian, 0-500m, weed
Use: 1. Cataracts, Impaired Eyesight / Whole plant, fresh / Topical / Break the Stems of the plant so that the juice drips out. Put milky sap in eye, 2 drops 3 times a day. **2.** Inflammation (external), Promoting lactation in women after birth, Helping the mother produce milk that will be accepted by the baby / Whole plant, fresh / Oral / Boil 10g Lecherita with 1 liter of water. Combine with 10g each of Cola de Caballo, Amor Seco, Linaza, and Chacur plus 5 Stems and 5 Leaves of Marrajudio. Drink 1 cup every other day for 4 days and/or wash with solution 2 times a day when needed.

Croton draconoides

Croton lechleri

Hura crepitans

Jatropha curcas

Manihot esculenta

Phyllanthus niruri

EUPHORBIACEAE - *Croton draconoides* Müll.-Arg.
Sangre de Grado, Sangre de Drago
Shrub, Amazonian, Andean, 0-2000m
Use: 1. Ulcers, Bleeding (Internal), Gastritis, Blood Circulation, Scars from insect bites, After internal surgery / Latex, fresh or dried / Oral / Cut only during the rainy season. Cut the Bark and allow 5 drops of the blood-like liquid into half a glass (8oz) of water. Patient should drink solution at room temperature, 3 times a day up to 2 months. Heals scars from the inside out. **2.** Wounds (external), Scars, Acne / Latex, fresh or dried / Topical / Cut the Bark and extract the resin that comes out of the Bark. Apply as a poultice. Use 3 drops that had turned into foam and rub on affected area, once a day as needed.

EUPHORBIACEAE - *Croton lechleri* Müll. Arg.
Sangre de Grado, Sangre de Drago
Tree, Amazonian, Andean, 0-1000m
Use: 1. Ulcers, Bleeding (Internal), Gastritis, Blood Circulation, Scars from insect bites, After internal surgery / Latex, fresh or dried / Oral / Cut only during the rainy season. Cut the Bark and allow 5 drops of the blood-like liquid into half a glass (8oz) of water. Patient should drink solution at room temperature, 3 times a day up to 2 months. Heals scars from the inside out. **2.** Wounds (external), Scars, Acne / Latex, fresh or dried / Topical / Cut the Bark and extract the resin that comes out of the Bark. Apply as a poultice. Use 3 drops that had turned into foam and rub on affected area, once a day as needed.

EUPHORBIACEAE - *Hura crepitans* L.
Habilla
Tree, Amazonian, Andean, 0-1500m
Use: Laxative, Overcoming Laziness / Seeds, dried / Oral / Grind 3 seeds and mix with 1 cup of oats. Make a hot cereal and drink warm, 1 cup once only. Drink cold. Do not exceed the dosage; it is very strong and can kill you.

EUPHORBIACEAE - *Jatropa curcas* L., *Jatropa gossypiifolia* **L.**, *Jatropa weberbaueri* Pax & Hoffman
Piñones
Shrub, Amazonian, Andean, Coastal, 0-1500m, weed
Use: Laxative, Overcoming Laziness / Seeds, dried / Oral / Grind 7 seeds and mix with 1 cup of oatmeal. Eat cereal warm. Patient will react with diarrhea and vomiting. After patient's reaction, serve a heavy tea. Drink 1 cup, once only.

EUPHORBIACEAE - *Manihot esculenta* Crantz
Yuca (Manihot)
Herb, Amazonian, Andean, Coastal, 0-2500m, cultivated
Use: 1. Vaginal Infection, Vaginal Discharge / Tuber, fresh / Oral / Boil 1 cup of water and add 50g of Yuca and boil for 5 minutes. Drink cold, 1/4 cup every day for 15 days. **2.** Allergies, Rashes / Tuber, fresh / Topical / Crush peel of the tuber and remove extract with a piece of cloth. Rub affected area with extract and leave on for 3 hours. After it is dried, bathe, 2 times a day for 20 days.

EUPHORBIACEAE - *Phyllanthus niruri* L., *Phyllanthus stipulatus* (Raf.) Webster, *Phyllanthus urinaria* L.
Chanca Piedra
Herb, Amazonian, Andean, 500-2500m
Use: Liver Inflammation, Blood Detoxification, Inflammation (Internal), Bladderstones, Liver, Kidneys, Gallbladder Inflammation / Whole plant, fresh or dried / Oral / Boil 10g of the plant, Cola de Caballo, Llantén, Boldo, Flor de Overo, Caña-Caña, Flor Blanca and Flor de Arena in 1 liter of water for 3-5 min. Drink 3 times a day for 2 weeks.

Ricinus communis

Acacia macracantha

Caesalpinia paipai

Caesalpinia spinosa

Cajanus cajan

Cassia fistula

EUPHORBIACEAE - *Ricinus communis* L.
Higrillo, Higrilla, Piñon, Higuerilla, Llonque
Shrub, Amazonian, Andean, Coastal, 0-2000m, weed, introduced
Use: 1. Constipation / Leaves, fresh / Topical / Put oil on the Leaf, then warm Leaf and Flowers over fire. Poultice: Place on the stomach and wrap in plastic. Apply hot or lukewarm. Do not ingest. Apply 2 times a day. **2.** Ulcers (External), Pimples, Wounds / Seeds, fresh / Topical / Grind 10g of seeds with 1/2 glass of Cañazo (Sugarcane Alcohol). Place mixture on top of the affected area. Leave it on for 2 hours, once only.

FABACEAE - *Acacia macracantha* Humb. & Bonpl. ex Willd.
Faique, Espino, Huarango
Tree, Amazonian, Andean, Coastal, 0-3000m, weed
Use: 1. Wounds, Stop Bleeding / Bark, dried / Topical / Burn Bark, collect ashes and strain. Place ashes on affected area. Cover entire wound, once a day until the wound is healed. **2.** Arthritis, Rheumatism / Bark, dried / Topical / Extract resin from 20g of Huarango. Dissolve resin with animal (snake, mule, chicken, guinea pig) fat. Place a small amount on the affected area. Massage as needed.

FABACEAE - *Caesalpinia paipai* Ruiz & Pav.
Pay Pay
Tree, Amazonian, Andean, Coastal, 0-2000m
Use: Killing Lice, Wounds / Fruit, fresh or dried / Topical / Boil 10 Fruits in 1 liter of water for 30 mins. Wash hair with cold wash, once a day in the AM for 3 days. Alternatively, apply once a day until wound heals. Amount applied depends on the size of the wound.

FABACEAE - *Caesalpinia spinosa (Molina)* Kuntze
Tara, Talla, Chanchalagua
Tree, Andean, Coastal, 0-3000m
Use: 1. Pharyngitis, Throat, Skin Infection, Animal Bites, Antibiotic, Tonsil Inflammation / Seed pods, fresh or dried / Oral / Boil 3 pods in 1 cup of water and mix with Romero, Coca, Fenegrew, Sangre de Grado and Vinegar. Gargle 3 times a day, morning and evening, Gargle and spit out, then drink 1 glass. Gargle and drink 1 cup in the morning, 1 at night for 6-7 months. Powdered seeds can be applied directly to Wounds. **2.** Fungus, Skin Infection, Angina Pectoris, Antibiotic, Wounds, Boils, Animal Bites, Amoeba Infections, Ovarian Inflammation, Uterine Inflammation, Vaginal Inflammation / Seed pods, fresh or dried / Topical / Boil 5g in 3 liters of water for 20 minutes, mixed with 10g each of Laurel, Hierba del Susto, Ajenco and Ishpingo. Bath: Once a week, or wash wounds 3 times a day for 1-3 months. Can also be used as a Vaginal Wash, 2 times a day for 2 days.

FABACEAE - *Cajanus cajan* (L.) Millsp.
Chivato
Shrub, Amazonian, Andean, 0-1500m, weed, introduced and cultivated
Use: Bad Air/Mal Aire / Whole plant, fresh or dried / Topical / Boil 10g total of Eucalyptus, Manzanilla, Ilambo, Cordón de Muerto, and Chivato in 3 liters of water for 5 minutes. Bathe patient with the water and rub with the plants, 2 times a week or 4 times a month as needed, depending on seriousness of the Bad Air/Mal Aire.

FABACEAE - *Cassia fistula* L.
Caña Fistula
Tree, Amazonian, Andean, 0-1000m, introduced and cultivated
Use: Nervous System, Epilepsy / Seeds, fresh or dried / Oral / Boil 10g in 1 liter of water. Drink 1 cup daily as needed (approximately 15-20 days).

Cicer arietinum

Desmodium molliculum

Desmodium triflorum

Dioclea virgata

Erythrina amazonica

Erythrina velutina

FABACEAE - *Cicer arietinum* L.
Garbanzo (Chickpea)
Herb, Amazonian, Coastal, 0-1000m, introduced and cultivated
Use: Cancer / Seeds, dried / Oral / Boil 1kg of Garbanzo in 1 liter of water for 5 minutes. Drink lukewarm, 1 cup a day for 15 days.

FABACEAE - *Desmodium molliculum* (H.B.K.) DC.
Pie de Perro, Pata de Perro, Chancas de Comida, Muña, Manayupa
Herb, Andean, 1000-3500m, weed
Use: 1. Inflammation (Internal and External), Kidney Inflammation, Diarrhea, Stomachache, Ovarian Inflammation, Gastritis / Whole plant, fresh or dried / Oral / Boil 10g of Pie de Perro in 1 liter of water. Combine with 10g each of Chacur, Amala, Amor Seco, and Verbena. Drink 4 times a day for 1 month. **2.** Wounds (Cleansing), Scars / Whole plant, fresh or dried / Topical / Boil with Llatén and Matico 20g total in 1 liter of water for 10 min. Wash once a day.

FABACEAE - *Desmodium triflorum* (L.) DC
Pega Pega
Herb, Amazonian, Andean, Coastal, 0-2500m, weed, introduced
Use: 1. Spiritual Flowering / Whole plant, fresh / Seguro / Standard Seguro mixture. **2.** Good Business, Protection, Good Fortune, Good Health / Whole plant, fresh / Topical / Standard mixture for Spiritual Flowering.

FABACEAE - *Dioclea virgata* (Rich.) Amsh.
Yin Yin
Vine, Amazonian, Andean, 0-1500m, weed
Use: 1. Guinea Pig Fertility / Whole plant, fresh / Oral / Serve the whole plant, 130g of plant every day. **2.** Promoting speech in children / Whole plant, fresh / Oral / Remove the seeds from the pod. Place the seeds in the child's mouth. Promptly instruct the child to close his/her mouth. Repeat 3 times. Repeat this procedure 2 times a day for 3 days.

FABACEAE - *Erythrina amazonica* Krukoff
Huayruro, Huairuro
Tree, Amazonian, 0-500m
Use: Protection from Evil / Seeds, dried / Amulet / Make a necklace with the seeds and have the shaman bless it. Wear the necklace for life.

FABACEAE - *Erythrina velutina* Willdenow
Porotillo
Tree, Andean, Coastal, 0-2500m
Use: Intestinal Cleansing / Flowers, Leaves and Stems, fresh or dried / Boil 5g in 1 liter of water. Adults: Add 2 tbsp of Glycerin Oil. Children: Add 5g Glycerin Oil; 1 enema every 3 months or 1 every 6 months depending on condition.

Indigofera suffruticosa

Inga edulis

Lathyrus odoratus

Lablab purpureus

Lens culinaris

Leucaena leucocephala

FABACEAE - *Indigofera suffruticosa* Miller
Añil
Herb or Shrub, Amazonian, Andean, 0-1500m, weed
Use: Expelling placenta from woman after giving birth / Stems, fresh / Oral / Boil 20g of Añil in 1 cup of water for 5 minutes. Drink lukewarm, 1/2 cup once only.

FABACEAE - *Inga edulis* C. Martius, *Inga feuilleei* DC
Guava, Huaba, Pacae, Pacai
Tree, Amazonian, Andean, 0-1500m, cultivated
Use: 1. Rehabilitation of drug addicts or alcoholics, Laxative / Seeds, fresh or dried / Oral / Grind 10 seeds and 10g total of Juan Alonso and Alcaparilla. Boil in 1/2 cup of water for 2 min. Mix with 1/2 glass of Orange Juice. Drink 3 - 4 times a day for 2 months or as needed. **2.** Adding shine and beauty to hair, Hair Growth / Flowers, fresh / Topical / Add 15g of the Flowers to 1 liter of water. Boil the mixture for 3 minutes. Let it sit. After shampooing, apply the mixture to the patient's hair without rinsing.

FABACEAE - *Lathyrus odoratus* L.
Tacón Blanco, Pensamiento Blanco
Herb, Andean, 2500-4000m, introduced and cultivated
Use: Heart, Nerves, Anxiety / Flowers, Leaves and Stems, fresh or dried / Oral / Boil 5g in 1 liter of water for 3 minutes combined with 10g each of Toronjil, Pimpinela, Mejorana and Cedrón. Drink 1 liter a day or 3-4 cups a day for 1 month.

FABACEAE - *Lablab purpureus* (L.) Sweet
Frijol Chileno
Shrub, Coastal, 0-1000m, weed, introduced and cultivated
Use: Fever, Intestinal Inflammation, Lung Protection / Fruits, fresh / Oral / Boil for 10 minutes 1/2kg of the plant material in 1 liter of water. Drink at room temperature, 1/2 cup 2 times a day for 8 days.

FABACEAE - *Lens culinaris* Medikus
Lenteja (Lentil)
Herb, Amazonian, Andean, Coastal, 0-3500m, introduced and cultivated
Use: Bone Protection / Seeds, dried / Oral / Boil 200g of Lenteja in 1 liter of water for 5 minutes. Drink cold, 2 times a day for 20 days. Eat Lentils with Rice.

FABACEAE - *Leucaena leucocephala* (Lam.) De Wit
Arabisca, Huaba Bruja
Tree, Amazonian, Andean, Coastal, 0-1500m, introduced and cultivated
Use: Antiseptic, Clean Wounds / Bark, Flowers and Stems, fresh / Topical / Boil 1 liter of water for 3-4 minutes with 10g total of Arabisca, Verbena, Hierba Santa, Llantén, Cola de Caballo and Chacur. Wash the Wound, Herpes or Rash 2 times a day or as needed until dry.

Lupinus mutabilis

Medicago sativa

Melilotus alba

Mimosa albida

Mimosa nothacacia

Myroxylon balsamum

FABACEAE - *Lupinus mutabilis* Sweet
Chocho, Tarhui
Herb, Andean, 1500-4500m, cultivated
Use: Malnutrition, Nutritional Supplement / Seeds, dried / Oral / Soak plant material in water for 1 week. Either puree the material with an onion and cook for 2 minutes or make a salad with the soaked material. Eat as a salad or as a puree. Eat 1 plate a day as necessary.

FABACEAE - *Medicago sativa* L.
Alfalfa
Herb, Andean, Coastal, 0-4500m, introduced and cultivated
Use: 1. Bronchitis / Flowers and Leaves, fresh / Oral / Blend Leaves and Flowers with water. Drain, and obtain extract. Drink extract. Honey can be added, if desired. Take 1 glass of extract twice a day. **2.** Bitterness in the Mouth, Kidney Disease / Flowers, fresh / Oral / Blend 20 Flowers in 1/4 cup of water. Drink 1 small cup once a month for 3 months. Do not exceed dosage or one might loose eyesight. Plant is very hot.

FABACEAE - *Melilotus alba* Medikus
Alfalfilla
Herb, Andean, Coastal, 0-4500m, introduced and cultivated
Use: 1. Gain weight / Seeds, dried / Oral / Boil 10g of ground seeds in 1/2 liter of water for 10 min. Drink lukewarm, 1/4 cup a day for 15 days. **2.** Fever, Tuberculosis, Colds, Respiratory Infections / Seeds, dried / Oral / Boil 100g of the plant material in 1/2 liter of water for 10 min. Drink cold, 1/2 a cup once a day for 8 days.

FABACEAE - *Mimosa albida* H. & B.
Tapa Tapa
Herb, Andean, Coastal, 0-2000m
Use: Hemorrage, especially after abortions. To prevent hemorrhages and heal the Uterus after giving birth. / Leaves, Stems, Flowers, fresh / Oral / Boil 10g of Tapa Tapa with some dark rock salt in 1/2 liter water for 5 minutes. Drink 1 cup twice a day for one week

FABACEAE - *Mimosa nothacacia* Barneby
Uña de Gato de la Costa
Shrub, Andean, 1000-1500m
Use: 1. Cancer, Kidney Inflammation, Hepatitis, Hemorrhoids, Liver Inflammation / Bark, dried / Oral / Boil 10g of the Bark in 1 liter of water for 3-4 minutes. Drink cool, 1 cup 3-4 times a day as needed. **2.** Anus Cyst, Vaginal Pimples, Anal Pimples / Bark, dried / Topical / Boil 200g of Uña de Gato de la Costa in 3 liters of water for 10 minutes and then fill a tub with the hot brew. Sit in it for 5 minutes, 2 times a week until the patient is cured.

FABACEAE - *Myroxylon balsamum* (L.) Harms.
Quina Quina, Kina Kina
Tree, Amazonian, 0-500m
Use: 1. Nervous System, Bad Air/Mal Aire, Epilepsy, Bronchitis / Seeds, dried / Oral / Grind 20 seeds mixed with 20 seeds each from 6 other plants: Ashango, Pucho, Amala, Ishpingo, Nuez Mozcada and Camalonga. Put in a bottle of wine and macerate for 8 days. Drink 3 small cups a day. **2.** Bad Air/Mal Aire, Bronchitis, Fright/Susto, Headache / Seeds, dried / Topical / Boil 20 seeds in 5 liters of water for 20-30 minutes. Combine with 20 seeds each of Ishpingo, Ashango, Pucho, Amala, Raucho, Tokio, Nuez Moscada, and Pepa de Cedrón macerated in 1 liter of 90 proof alcohol. Add 2 pieces of tobacco, 2 pieces of Ajo Macho, 10g of Quina Quina, 2 Leaves of Pacra and 1 branch each of Eucalyptus and Maye. Do not leave bath outside. Take bath every other day, 3 times a week. **3.** Cough, Bronchitis, Asthma / Seeds, dried / Oral / Toast and crush 3 seeds in 1 cup of water. Drink 1/2 cup for adults, 1 tsp for children.

Otholobium mexicanum

Ormosia sp.

Pisum sativum

Prosopis pallida

Senna alexandrina

Senna bicapsularis

FABACEAE - *Otholobium mexicanum* (L.f.) Grimes
Culén
Shrub, Andean, 1000-4000m
Use: Diarrhea, Cold in the Stomach, Diabetes / Stems, fresh or dried / Oral / Boil 5g in 1 liter of water. Combine with 10g each of Manzanilla, Menta, and Anis. Drink 3 times a day. Patient should drink warm solution.

FABACEAE - *Ormosia* sp.
Huayruro, Huairuro
Tree, Amazonian, 0-500m
Use: Evil Eye/Mal Ojo / Seeds, dried / Amulet / Make a bracelet with the Seeds. Wear at all times on the left hand

FABACEAE - *Pisum sativum* L.
Arbejas, Arvejas (Pea)
Herb, Andean, Coastal, 0-4500m, introduced and cultivated
Use: Smallpox, Rubiola, Intestinal Inflammation, Release of all negativity, / Seeds, fresh / Oral / Boil 100g in 1 cup of water. Drink warm, 1 cup once or twice for 1 day only. Patient should drink cold solution, 1/2 cup once only.

FABACEAE - *Prosopis pallida* (H. & B. ex Willd.) H.B.K.
Algarrobo (Carob)
Tree, Andean, Coastal, 0-1500m
Use: 1. Cough, Anaemia, Fertility, Sexual Potency, Bronchitis, Nutritional Supplement / Seeds, dried / Oral / Boil 10kg of Algarrobo Fruit and Seeds for 3 hours in medium to high heat until thickened. Turn off fire and let sit until cool. Drain and place syrup in bottle. Drink 2 tbsp in 1 small cup, 3 times a day as long as you wish. **2.** Stomachache, Hangover / Leaves and Stems, fresh / Oral / Boil 5g of Algarrobo Bark in 1/4 cup of water for 3 minutes. Drink. **3.** Arthritis, Rheumatism, Colds, Bone Pain / Resin, fresh / Topical / With a knife extract the Resin exposed on the trunk. Place 5g of Resin in a pan to warm. Massage the affected area with the Resin, once a week for 3 weeks. **4.** Toothache, Tooth Extraction / Resin, fresh / Topical / Place 2 drops on top of tooth cavity. The Algarrobo Resin will pulverize the tooth. Patient should be very careful while applying because it will destroy all teeth touched by the Resin. **5.** Critical Wounds / Resin, fresh / Topical / Grind 100g of Algarrobo Charcoal, 100g of sulphur, and 100g of garlic. Apply on affected area, once a day until healed.

FABACEAE - *Senna alexandrina* Mill.
Hojas de Sen
Herb, Coastal, 0-1500m, introduced and cultivated
Use: Purgative, Constipation, Cleansing of the Stomach / Leaves, dried / Oral / Boil 3g in 1 glass of water for children, 5g per glass of water for adults, once a month.

FABACEAE - *Senna bicapsularis* (L.) Roxburgh
Alcaparrilla, Alpacaquilla
Herb, Coastal, 0-1000m, weed
Use: Alcohol and Drug Detoxification, Liver and Kidney Detoxifiction / Whole plant, fresh / Topical / Add 10g of Alcaparrilla to 1 liter of boiling water. Also add 30g total of a mixture consisting of Cola de Caballo, Juan Alonso, and Verbena. Boil the mixture for 3-5 min. Drink 1 cup, 3-4 times a day for one month or as needed.

Senna occidentalis

Spartium junceum

Tamarindus indica

Trifolium repens

Zornia piurensis

Coutoubea ramosa

FABACEAE - *Senna occidentalis* (L.) Link.
Retama, Retania
Herb, Amazonian, Andean, 0-1000m, weed
Use: Inflammation (general) / Whole plant, fresh or dried / Oral / Boil 10g total of Retama, Amor Seco, Cola de Caballo, Linaza, Chacur, and Pie de Perro in 1 liter of water. Drink warm, 1 cup 4 times a day for 1 month.

FABACEAE - *Spartium junceum* L.
Retama
Shrub, Andean, Coastal, 0-4000m, weed, introduced
Use: 1. Hepatitis, Liver, High Blood Pressure, Diabetes / Flowers and Root, fresh / Oral / Boil 3-5g in 1 liter of water combined with 3-5g of Flor de Overo. Drink 3 times a day. **2.** Arthritis, Good Luck, Bone Pain, Sinusitis, Blood Purification / Whole plant, fresh / Topical / Boil for 20 min. and mix with Maique. Bathe 3 times a week for one week, 50g total boiled 20-30 min. in 5 liters of water with herbs of luck and strength (Condor and Trensilla) or as Steam Bath, 3 times a week.

FABACEAE - *Tamarindus indica* L.
Tamarindo
Herb, Amazonian, Coastal, 0-500m, weed, introduced and cultivated
Use: Laxative, Blood Circulation, Epilepsy, Heart Disease / Fruit pulp, fresh / Oral / Remove the pulp from 250g of plant material. Add this to 3 glasses of warm water. Blend the mixture. Drink the mixture cold while fasting, 1 glass in the morning, once a day for 30 days. Repeat as necessary.

FABACEAE - *Trifolium repens* L.
Trebol, Trebol de Agua (Clover)
Herb, Andean, Coastal, 0-4500m, weed, introduced
Use: Inflammation, Urinary Tract Inflammation, Stomach, Stomach Pain, Ulcer, Kidney Inflammation, Blood Enrichment / Flowers, Leaves and Stems, fresh or dried / Oral / Boil 5g in 3 liters of water. May combine with Lancetilla and Colores, 5g each. May also sometimes combine with Zarzaparilla (1 tbsp). Drink 1 liter daily for 1-2 months or take 5g of the freshly chopped plant in the morning every day for 1 week.

FABACEAE - *Zornia piurensis* Mohlenbrock
Hierba de la Víbora
Herb, Coastal, 0-1000m
Use: Nervousness / Whole plant, dried / Oral / Boil 5g total in 1 liter of water with Canchalagua, Nabo, and Colores. Drink 3 times a day during meals.

GENTIANACEAE - *Coutoubea ramosa* Aubl.
Genciana
Herb, Amazonian, 0-500m, weed
Use: Wounds, Scars, Rashes, Cold Sores / Fruits, fresh / Topical / Squeeze the juice out of the Fruit. Apply 2 drops of the fresh squeezed juice on the affected area and let it dry naturally. Apply 2 drops once a day for as long as needed.

Gentianella bicolor

Gentianella brunneotincta

Gentianella crassicaulis

Gentianella dianthoides

Gentianella graminea

Erodium cicutarium

GENTIANACEAE - *Gentianella bicolor* (Wedd.) Fabris ex J.S.Pringle
Corpus Way, Hórnamo León
Herb, Andean, 3000-4500m
Use: 1. Arthritis, Diabetes, Bone Pain, Cholesterol, Gastritis, Liver, Blood, Rheumatism / Whole plant, fresh or dried / Oral / Boil 2-3 minutes in 1 liter of water. Drink daily as needed. Tea is very bitter. **2.** Good Luck, Good Health, Good Business, Protection / Whole plant, fresh or dried / Seguro / Mix plant material with Hierba de la Plata, Hierba de la Fortuna, Hierba del Dolar, Hierba de la Justicia, Hierba del Oro, Carpintero, Señorita and Sonrisa. Place all the herbs in 1 bottle with Agua Florida and Lime Juice. Use mixture for rituals or as a Seguro.

GENTIANACEAE - *Gentianella brunneotincta* (Gilg) J.S.Pringle
Anga Macha
Herb, Andean, 3500-4500m
Use: Uterine Infection after giving birth / Whole plant, fresh / Oral / Boil 1/2 liter of water with 5g of Valeriana Estrella and 5g of Anga Macha for 3 min. Drink hot, 1 glass 3 times a day for 2-3 days.

GENTIANACEAE - *Gentianella crassicaulis* J.S.Pringle
Violeta Genciana
Herb, Andean, 3000-4500m
Use: Gastritis, Diabetes, Dizziness / Whole plant, fresh or dried / Oral / Boil 30g in 1 liter of water for 3-5 minutes. Combine with 10g each of Pasuchaca, Amargón, and Corpus Way. Drink 3-4 glasses a day for 15-30 days.

GENTIANACEAE - *Gentianella dianthoides* (H.B.K.) Fabris
Genciana, Egenciana, Amargón, Campanilla
Herb, Andean, 3000-4500m
Use: Liver, Kidneys, Blood, Purgative, Diabetes, Cleansing, Blood Irrigation, Blood problems (general), Liver Infection / Whole plant, fresh / Oral / Boil 10g of Genciana in 1 liter of water for 2 minutes. Drink once a day before eating in the evening, every other day for 1 week. Overdosing can cause miscarriage in pregnant women. The plant contains cortizone.

GENTIANACEAE - *Gentianella graminea* (H.B.K.) Fabris
Sumarán, Chinchimali, Corpushuay
Herb, Andean, 2500-4500m
Use: Diabetes, Liver, Blood, Fat Burner, Intestinal Fever, Cough, Fever, Infection, Blood Allergies, Varicose Veins, Blood Purification, Liver Inflammation, Blood Detoxification / Whole plant, fresh or dried / Oral / Boil 20g in 1 liter of water. Drink 1 liter daily for 1 week. It is best taken with food since it has a bitter taste. Drink cool while the patient is fasting. Exceeding dosage can lead to blindness.

GERANIACEAE - *Erodium cicutarium* (L.) L'Herit.
Agujilla Blanca, Auguilla.
Herb, Andean, Coastal, 500-4500m, weed
Use: Inflammation (General), Bronchitis, High Blood Pressure, Low Blood Pressure / Whole plant, fresh / Oral / Boil 5g of the sap in 1 liter of water mixed with 5g each of Ambarindas, Hierba del Toro and Sanguinaria. Drink 1 liter a day for 1-3 months.

Geranium ayavacense

Pelargonium odoratissimum

Pelargonium roseum

Hypericum aciculare

Hypericum laricifolium

Hypericum silenoides

GERANIACEAE - *Geranium ayavacense* Willd ex H.B.K., *Geranium sessiliflorum* Cavanilles
Puli Punchi, Pasuchaca, Pachuchaca, Miscamisca
Herb, Andean, 3000-4500m
Use: 1. Diabetes / Whole plant, fresh or dried / Oral / Boil 1 liter of water for 3 minutes. Add 10g of Pasuchaca and 1/2 Leaf of Nogal. Combine with 10g each of Culén and Citrodora. Drink 4 cups a day for life. **2.** Inflammation, Kidneys, Liver, Urinary Tract, Inflammation (General) / Whole plant, fresh or dried / Oral / Boil 1 liter of water with 10g of Pasuchaca for 3 minutes. Combine with 10g each of Chacur, Cola de Caballo, Verbena, Unquia, Amor Seco, and Grama Dulce. Drink 4 cups a day for 1 month.

GERANIACEAE - *Pelargonium odoratissimum* (L.) L'Herit.
Malva de Oro, Malva Olorosa.
Herb, Andean, 1500-3500m, introduced
Use: 1. Arthritis, Heart, Nerves, Blood, Fright/Susto, Ovarian Inflammation, Uterine Inflammation / Whole plant, fresh or dried / Oral / Boil 5g in 1 liter of water. Drink 1 liter a day. **2.** Fright/Susto / Whole plant, fresh or dried / Topical / Boil 2 liters of water with 10g each of Ishpingo, Eucalyptus, Cordón del Muerto, Flor de Chocho, and Flor de Retama. Boil for 3 minutes. Bathe 2-3 times a month.

GERANIACEAE - *Pelargonium roseum* Willd.
Geranio (Geranium)
Herb, Andean, 1500-3500m, introduced
Use: Hemorrhages, Uterine Pain, Uterine Inflammation, Tonsillitis, Throat Infection / Flowers and Leaves, fresh / Oral / Boil 10g of Geranio in 1 liter of water. Drink 3 times a day as needed. Solution also can be used to gargle, 3-4 times daily for 3-4 days.

HYPERICACEAE - *Hypericum aciculare* Kunth.
Hierba de las Cordilleras, Lechuguilla, Hierba del Imán
Shrub, Andean, 2000-3500m
Use: 1. To assure realization of one's needs and wants / Leaves and Stems, fresh / Seguro / 1 small stem. **2.** Daño/Sorcery, Headache / Leaves and Stems, fresh / Topical / Mix with Sugarcane Alcohol, 7 Espiritus and Hierba Santa. Limpa: 2 times a week. **3.** Fever (General), Intestinal Fever / Leaves and Stems, fresh / Oral / 3 Leaves chopped and made into extract. No mixing! 5g a day for 8 days. **4.** Daño/Sorcery, Headache / Leaves and Stems, fresh / Topical / Natural, with 7 Espiritus. Apply as Poultice for 4 hours.

HYPERICACEAE - *Hypericum laricifolium* Juss.
Hierba del Cariño, Hierba de la Fortuna, Solitario, Chinchango, Abrecaminos
Shrub, Andean, 2500-4500m
Use: 1. Fragrance, Luck in Love and Work, Bad Air/Mal Aire, Protection, Good Fortune, Good Health / Whole plant, fresh / Topical / Alternative mixture for Spiritual Flowering. Take 3 baths a month. **2.** Good Business, Protection, Good Fortune, Good Health / Whole plant, fresh / Seguro / Standard Seguro mixture.

HYPERICACEAE - *Hypericum silenoides* Juss.
Cintaura
Herb, Andean, Coastal, 500-3500m
Use: Diarrhea, Dysentery / Whole plant, fresh / Oral / Combine 3-5g of Cintaura in 1 liter of water mixed with 10g each of Culen and Hierba del Toro. Drink 3 times a day for 5 days or as needed.

Hesperoxiphion niveum

Isoetes andina

Juglans neotropica

Krameria lappacea

Clerodendrum philippinum

Hyptis sidifolia

IRIDACEAE - *Hesperoxiphion niveum* (Rav.) Rav.
Hierba de la Justicia, Piri Piri, Totorilla
Herb, Andean, 1500-2500m
Use: 1. House Protection, Health Protection, Professionl Success, Wounds, Fragrance, Good Luck, Love, Domination in matters of Justice (ritual), Domination in legal matters (ritual) / Whole plant, fresh / Seguro / Put inside a bottle 10g of Hierba de la Justicia and add Hierba de la Plata, Dolar, Fortuna, Señorita, Valeriana Estrellada, Tabú Perfume, Agua Florida, Lime Juice, Sugar, and Agua Bendita. Keep Seguro in bedroom. As alternative put together in a cloth 10g of Hierba de la Justicia, 10g of Hierba del Dominio and 10g of Hierba del Olvido. Seal and pray. Patient must carry the bag on his or her person and use it with prayers. **2.** Daño/Sorcery via the mouth, Daño/Sorcery (General), Purgative / Whole plant, fresh / Oral / Have patient eat 3 fresh plants. Patient must be kept out of the light, staying inside for 1 week. Will probably cause vomiting. After consumption patient has to follow a strict diet of no spices at all for 1 week **3.** Spiritual Flowering / Whole plant, fresh / Topical / Boil Hierba de la Justicia in some water for 3-4 minutes. Bathe 3-4 times a month.

ISOETACEAE - *Isoetes andina* Spruce ex Hook.
Piri Piri
Herb, Andean, 3000-4500m
Use: Male Impotence / Stems, fresh / Oral / Boil 2 small branches in 1 cup of water. Drink 1 cup a night for 1 month.

JUGLANDACEAE - *Juglans neotropica* Diels
Nogal (Walnut)
Tree, Amazonian, Andean, 0-3000m
Use: 1. Hair Loss / Leaves, fresh / Topical / Boil 20g in 1 liter of water for 20 minutes. Massage head for 3 minutes, 3 times a week. **2.** Daño/Sorcery, Arthritis, Wounds (Cleansing), Fright/Susto / Leaves, fresh / Topical / Boil 3 liters of water with a bundle or pouch of the herb, 3 times a month. **3.** Cough, Bronchitis, Asthma / Leaves, fresh / Oral / Boil 10g in 1 liter of water for 3-5 minutes. For Bronchitis: Mix with Matico, Enredadera and Borraja, 10g each. Drink 3 glasses a day, 1 liter daily. **4.** Diabetes / Leaves, fresh / Oral / 5g with 5g of Pasuchaca, 1 liter daily. **5.** Daño/Sorcery, Arthritis, Wounds (Cleansing), Fright/Susto / Leaves, fresh / Topical / Limpia: Mix with Añasquero Grande, Rumilanche, Ishpinguillo and Sauco, 3 times a month.

KRAMERIACEAE - *Krameria lappacea* (Dombey) Burdet & B.B. Simpson
Ratania
Shrub, Andean, Coastal, 500-4000m
Use: Kidney Inflammation, Ovarian Inflammation, Intestinal Inflammation, Inflammation of the Bladder / Leaves and Root, fresh / Oral / Boil 10g in 1 liter of water for 3 minutes. Drink as needed.

LAMIACEAE - *Clerodendrum philippinum* Schauer
Brochamelia
Shrub, Amazonian, Andean, 0-1000m, introduced and cultivated
Use: Bronchitis, Asthma, Whooping Cough, Stenosing laryngotracheitis (pseudo-croup, a condition, where a baby can't breathe, turns blue, and makes a "rooster like" noise) / Flowers, fresh or dried / Oral / Boil 1 liter of water and add 10g of the herb. Boil for 3 to 5 minutes more. Can be mixed with 10g each of Huamanripa and Veronica. Drink 1 cup 3 times a day for 2 weeks.

LAMIACEAE - *Hyptis sidifolia* (L'Her.) Briq.
Pedrorera, Hierba de la Ventosidad, Albaca Serrana
Herb, Andean, Coastal, 0-3500m, weed
Use: Gases, Intestinal Colic, Gastritis, Cramps / Whole plant, fresh or dried / Oral / Boil 5g in 1 liter of water for 3-5 minutes. Drink 1 cup, 3-4 times a day for 2-3 days as needed. When stomach is hard, this plant loosens it and gases are released.

Lavandula angustifolia

Lepechinia meyenii

Marrubium vulgare

Melissa officinalis

Mentha piperita

Mentha spicata

LAMIACEAE - *Lavandula angustifolia* Miller
Alucema, Labanda
Herb, Andean, Coastal, 0-1500m, introduced and cultivated
Use: 1. Cold, Gases, Heart, Nerves / Flowers, Leaves, Stems and Seeds, dried / Oral / Do not use roots. Boil 1 liter of water. Add a total of 10g of Alucema, Romero, Claveles, Hinojo, Toronjil, Anjenco, Manzanilla and Pimpinela. Boil for 2 min more. Patient should drink lukewarm solution, 1 cup 3-4 times a day for 1 month. **2.** Relaxant, Stress / Flowers, Leaves, Stems and Seeds, dried / Topical / Boil 3 liters of water with 20g total of Alucema, Romero, and Eucalypto for 3 minutes. Patient should wash with water and massage with Flowers, 2 times a week or 4 times a month or as needed.

LAMIACEAE - *Lepechinia meyenii* (Walp.) Epling
Salvia, Salvia Real
Herb, Andean, 2500-4500m, weed
Use: 1. Bronchitis, Heart, Nerves, Memory, Menstruation / Whole plant, fresh or dried / Oral / Boil 30g in 1 liter of water. Drink with meals, 3 times a day. **2.** Wounds, Hair loss / Whole plant, fresh or dried / Topical / Boil 100g total in 8 liters of water for 5 minutes combined with Romero and Llantén. Bathe 3 times a day for 1 month. **3.** Fright/Susto / Whole plant, fresh or dried / Topical / Boil 5g in 3 liters of water mixed with Añasquero Chico, Ajenco, Nogal and Vinegar (5g each). Bathe 2 times a month.

LAMIACEAE - *Marrubium vulgare* L.
Cordón del Muerto, Chanca de Comida, Chancas del Muerto
Herb, Andean, Coastal, 500-4000m, introduced
Use: Fright/Susto, Inflammation (external) / Whole plant, fresh or dried / Topical / Boil 20-30g total of Cordon del Muerto mixed with Eucalyptus, Ishpingo, Chivato, and Quinual in 2-3 liters of water for 5-8 minutes. Empty into a bucket and place in a closed room. Rub the solution on your whole body with all the plants, 2 Baths a week for 4-5 months if the patient is in good condition. Do not let the solution touch the mouth.

LAMIACEAE - *Melissa officinalis* L.
Toronjil, Melissa
Herb, Andean, Coastal, 0-3500m, introduced
Use: 1. Ill-mannered children / Whole plant, fresh or dried / Topical / Boil 10g in 2 liters of water for 10 minutes. Mix with 10g each of Toronjil and Churguis. Bathe twice a week or as needed. **2.** Lovesickness, Nerves, Insomnia, Heart, Nervous System, Tachycardia / Whole plant, fresh or dried / Oral / Boil 20-30g in 1 liter of water for 2 minutes with 10g each of Pimpinela, Cedrón, Mejorana, Siempre Viva, Flor de Amelas, Romero, Claveles, Congona, Manzanilla, Mejorana, and Orange Flowers. Drink 4 cups a day for 1 month.

LAMIACEAE - *Mentha piperita* L.
Poleo (Peppermint)
Herb, Andean, Coastal, 0-3500m, inroduced
Use: Colic, Stomachache / Whole plant, fresh or dried / Oral / Boil in 1 liter of water. Add 10g of Poleo. Take when symptoms occur.

LAMIACEAE - *Mentha spicata* L.
Hierba Buena, Hierba Buena Silvestre, Menta (Mint)
Herb, Andean, Coastal, 0-3500m, introduced
Use: 1. Parasites, Colic, Stomachache, Gastritis, Indigestion, Tapeworms, Intestinal Worms, Headache, Aphrodisiac, Gases, Bad Breath / Whole plant, fresh / Oral / Boil 10g in 1 liter of water. Can be mixed with 10g of Anís. Drink 3 times a day after each meal for 1 month. **2.** Colic, Stomachache / Whole plant, fresh / Topical / Boil for 20 minutes, 30g in 6 liters of water with other Good Luck herbs (10g each). Bathe 3 times a week.

Minthostachys mollis

Ocimum basilicum

Origanum majorana

Origanum vulgare

Rosmarinus officinalis

Salvia ayavacensis

LAMIACEAE - *Minthostachys mollis* (Benth.) Griseb.
Muña, Chancas de Comida
Herb, Andean, Coastal, 500-3500m, weed
Use: Colic, Gases, Stomach Parasites, Stomach Ache, Heart, Nerves, Diarrhea / Leaves and Stems, fresh / Oral / Boil 2-10g in 1 liter of water. Drink 1 cup 3-4 times a day for 3-4 weeks. Patient should drink hot solution. Can also be eaten as a vegetable or in soup.

LAMIACEAE - *Ocimum basilicum* L.
Albaca, Albaca Mixtura, Albaca Negra, Albaca Morada (Basil)
Herb, Andean, Coastal, 0-3500m, introduced and cultivated
Use: 1. Daño/Sorcery, Fright/Susto, Bad Air/Mal Aire, Insomnia, Low Blood Pressure, Good Luck / Whole plant, fresh / Whole plant, fresh / Use 20g of the herb mixed with Agua Florida, Eucalyptus, Camphor, Molle, Sugarcane Alcohol, Agua del Susto, Flor de Retama, Hierba del Gallinazo, Ruda (Hembra and Macho), Ajenco, 2 Chilis (20g of each). Limpia: Perform on Tuesday, Friday, and Tuesday. **2.** Uterus Dilation to hasten delivery, Prevention of Birth-related Infections, Refresh the Womb, Reduce After-birth Inflammations, Colic, Gases / Whole plant, fresh / Oral / Boil 10g in 1 cup of water with a little salt. Drink 1 cup, very hot. Drink also immediately after giving birth. **3.** After birth / Whole plant, fresh / Oral / Boil 5g in 1 liter of water. Drink 1 cup a day for 2 weeks. **4.** Daño/Sorcery, Fright/Susto, Bad Air/Mal Aire, Insomnia, Low Blood Pressure, Good Luck / Whole plant, fresh / Oral / Boil 20g in 1 liter of water for 1-2 minutes combined with Tilo, Toronjil, Mejorana, and Cedrón. Drink 3 cups a day before dinner for 10 days. **5.** Daño/Sorcery, Fright/Susto, Bad Air/Mal Aire, Insomnia, Low Blood Pressure, Good Luck / Whole plant, fresh / Whole plant, fresh / Boil 10g in 1 liter of water for 5-8 minutes combined with Romero, Salvia Real, Yerba Santa, and Malva Olorosa (10g each). Bathe 3 days a week or every 8 days for 2 months. **6.** Cataracts, Eye Problems / Whole plant, fresh / Whole plant, fresh / Place 1 seed directly into the eye. Leave for 3 hours or until the eye flushes it out. Apply once a month for 2 months.

LAMIACEAE - *Origanum majorana* L.
Mejorana
Herb, Andean, Coastal, 0-3500m, introduced and cultivated
Use: Colic, Heart, Nerves, Menstruation, Anxiety, Depression, Pain of love / Leaves and Stems, fresh / Oral / Boil 10g of Mejorana in 1 liter of water combined with Sanguinaria, Congona, Toronjil, Hinojo, Albahaca Serrana, Poleo, and Manzanilla (10g each). Patient should drink warm solution, 4 cups a day for 1 month. For Diabetes: 4 times a day for life.

LAMIACEAE - *Origanum vulgare* L.
Oregano
Herb, Andean, Coastal, 0-3500m, introduced and cultivated
Use: Colic, Menstrual Cramps, Menstruation, Stomachache, Gases, Stomach Cramps related to PMS / Leaves and Stems, fresh or dried / Oral / Add 1 cup of water to 3g of Oregano. Boil for 3-5 minutes. Drink hot for Menstrual Periods. Drink warm for Colic. Avoid drinking excessive amounts which may cause abortion. Drink 2 cups for 1 day only.

LAMIACEAE - *Rosmarinus officinalis* L.
Romero, Romero Castilla (Rosemary)
Herb, Andean, Coastal, 0-3500m, introduced and cultivated
Use: 1. Hair Loss / Leaves, fresh or dried / Topical / Boil 5g each of Romero, Nogal, Amor Seco, and Cola de Caballo in 1 liter of water for 5 min. Massage scalp with some of the prepared liquid. Apply shampoo and wash. **2.** Gases, Heart, Nerves, Bronchitis, Indigestion, Colic, Headache, Stomach, Lovesickness / Leaves, fresh or dried / Oral / Boil 5g each of Eucalypto, Cola de Caballo, and Ruda in 1 liter of water. Drink 2 cups a day for 2 days, 4 cups a day for 1 month. **3.** Fright/Susto, Expelling Negative Energy from the house / Leaves, fresh or dried / Topical / Boil 1 bundle of Romero with 3 liters of water for 10 minutes. Combine with 10g each of Llantén, Manzanilla Blanca, Sávila, Palo Blanco, Manzanillón, Retama Flowers, Chochos Flowers, Cordón del Muerto, Verbena, Toronjil, Eucalyptus and 7 Espiritus. Inhale or bathe Tuesday, Friday and the following Tuesday for 10 minutes each time. **4.** Purify, Absorb Negative Energy, Humidity Reduction / Leaves, fresh or dried / Topical / Burn 1 bundle of Romero and Palo Santo on top of burning charcoal. Inhale.

LAMIACEAE - *Salvia ayavacensis* Kunth
Ticra
Herb, Andean, 2500-3500m
Use: Body Fungus, Daño/Sorcery, Fright/Susto / Leaves, fresh or dried / Topical / Put 1 handful in 1 liter of boiling water. May combine with Tutapure Blanco, Lailambo, Conchalay Amarillo, Conchalay Blanco,

Salvia cuspidata

Salvia discolor

Salvia macrophylla

Salvia officinalis

Salvia officinalis subsp. *lavandulifolia*

Salvia sagitatta

Ticra, Zanahoria, Poleo de Gentil (5g each), and 7 Espiritus. Bathe 1-2 times a month. Do not ingest.

LAMIACEAE - *Salvia cuspidata* R. & P.
Salvia Blanca
Herb, Andean, 2000-3500m, weed
Use: Fright/Susto, Daño/Sorcery / Whole plant, fresh or dried / Topical / 1 bundle boiled for 5 minutes in 3 liters of water. Bathe once a week.

LAMIACEAE - *Salvia discolor* H.B.K.
Palmeras (Chica), Llatama, Yatama
Herb, Andean, 1000-2500m, weed
Use: 1. Success / Stems, fresh / Seguro / Place 3 Stems in 1 Flask. **2.** Cough / Stems, fresh / Oral / Boil 3 Leaves in 1 cup of water. Do not mix with other herbs. Drink 1 cup a day for 1 week. **3.** Clean the energy of the home, Prevention of Birth-related Infections, Fright/Susto in children / Stems, fresh / Topical / Used with Ishpinguillo, Hierba del Gallinazo, Ajosquiro and 7 Espiritus. Use for Steam Bath twice a month for adults or Bath 2 times a month for children. **4.** Prevention of Birth-related Infections / Stems, fresh / Oral / Boil 5g in 1 cup of water. Drink 1 cup daily for 1 month.

LAMIACEAE - *Salvia macrophylla* Benth.
Cuchichara
Herb, Andean, Coastal, 0-3500m, weed
Use: Sorcery Wounds, Gangrene / Leaves, fresh or dried / Topical / Toast and beat into a pulp. Put pulp on Wound.

LAMIACEAE - *Salvia officinalis* L.
Salvia (Sage)
Herb, Andean, Coastal, 0-3500m, introduced and cultivated
Use: Cough, Bronchitis, Menstrual Cycle Regulation / Whole plant, fresh or dried / Oral / Boil 10g of the plant in 1 liter of water for 3-5 minutes. Can be mixed with Matico, Nogal, and Eucalypto (10 g each). Drink hot, 1 cup 3-4 times a day as needed, up to one month.

LAMIACEAE - *Salvia officinalis* subsp. *lavandulifolia* (Vahl) Gams
Romero del Campo, Romero Blanco, Romero Serrano
Herb, Andean, Coastal, 0-3500m, introduced and cultivated
Use: 1. Inflammation, Bad Air/Mal Aire, Negative Energy, Cleansing of the Home, Fright/Susto, Sinusitis / Whole plant, fresh / Topical / Boil 20g total in 1 liter of water for 20 min with other Romeros, Salvia, and Mejorana. Administer while tepid and absorb the vapors produced by steam inhalation 2-3 times a month. **2.** Inflammation, Bad Air/Mal Aire, Negative Energy, Cleansing of the Home, Fright/Susto, Sinusitis / Whole plant, fresh / Topical / Alternatively burn one bundle with Charcoal, Palo Santo, Incense, Myrrh, and Eucalypto. Patients should surround themselves with the vapors and smoke produced. **3.** Inflammation, Bad Air/Mal Aire, Negative Energy, Cleansing of the Home, Fright/Susto, Sinusitis / Whole plant, fresh / Topical / Bath: Boil 1 bundle or pouch each in 3 liters of water with Canchalagua, and Cola de Caballo. When applied fresh, use Leaf and Stems; dried: use only the Leaf. Apply once only (Tuesday or Friday).

LAMIACEAE - *Salvia sagitatta* Ruiz & Pav.
Salvia Negra
Shrub, Andean, 2500-3500m, weed
Use: Cough, Asthma, Hair Loss / Root and Stems, fresh or dried / Oral / Boil 10g in 1 liter of water. Drink 3 times a day as needed

Salvia tubiflora

Satureja pulchella

Scutellatia scutellarioides

Stachys lanata

Thymus vulgaris

Aiouea dubia

LAMIACEAE - *Salvia tubiflora* R. & P.
Hierba de la Recaida, Hierba del Aire, Cutiquero, Yuca del Aire, Paja del Aire
Herb, Andean, Coastal, 0-3500m, weed
Use: 1. Bad Air/Mal Aire, Body Pain / Whole plant, fresh or dried / Oral / Boil 5g in 1 liter of water. Do not mix. Drink 1 cup 3 times a day. **2.** Bad Air/Mal Aire, Aneurysm, Paralysis, Body Pain / Whole plant, fresh or dried / Topical / Boil 50g in 1 liter of Sugarcane Alcohol for 20 minutes. Add 1 bottle of Agua Florida, 1 bottle of Agua Cananga and finish filling the bottle with Sugarcane Alcohol. Bathe 3 times a week.

LAMIACEAE - *Satureja pulchella* (Kunth) Briquet
Panisara
Shrub, Andean, 2500-3500m, weed
Use: Bronchitis, Asthma, Liver Disease, Infection (Internal), Nerves, Menstrual Delay, Vitamin / Leaves, fresh or dried / Oral / Boil 50g total of Panisara, Culein, Manzanilla and Chancas de Comida in 1/2 cup of water for 3 min. Drink the mixture cold. Take 1/8 cup once a day for 3 days.

LAMIACEAE - *Scutellatia scutellarioides* (Kunth) R. Harley
Poleo de Gentil
Herb, Andean, 1000-3500m
Use: Fright/Susto, Daño/Sorcery / Whole plant, fresh / Topical / Boil 1 bundle for 5 min in 3 liters of water for 1 bath.

LAMIACEAE - *Stachys lanata* Jacq.
Veronica (Macho)
Herb, Andean, 2000-3500m, introduced
Use: Bronchitis, Asthma / Whole plant, dried / Oral / Boil 10g of Veronica Macho with 1 liter of water for 5 min. Combine with Salvia, Matico, and Muyaca (10g each). Drink after meals, 3 cups a day for 15 days.

LAMIACEAE - *Thymus vulgaris* L.
Tomillo (Thyme)
Herb, Andean, Coastal, 0-3500m, introduced and cultivated
Use: Cough, Colic, Liver, Gases, Indigestion, Bladder / Leaves, Stems, and Flowers, fresh or dried / Oral / Boil 5g in 1 liter of water for 3 min. Drink 3 times a day.

LAURACEAE - *Aiouea dubia* (Kunth) Mez
Ishpingo
Tree, Andean, 500-2000m
Use: 1. Fright/Susto, Bad Air/Mal Aire / Seeds, fresh or dried / Topical / Crush and boil 20 seeds in 5 liters of water for 20-30 min. Mix with Ishpingo, Ashango, Pucho, and Amala (20 seeds each). Bathe 3 times a week. Do not leave bath outside. **2.** Bad Air/Mal Aire, Epilepsy / Seeds, fresh or dried / Oral / Crush seeds and mix with seeds of 6 other plants: Ashango, Pucho, Amala, Quina Quina, Nuez Moscada, and Camalonga. Boil in water for 20-30 min. Drink once a month. **3.** Rheumatism / Seeds, fresh or dried / Topical / Boil 5 seeds in 1/2 liter of water or Agua Florida. Rub affected areas once a day.

Cinnamonum verum

Nectandra reticulata

Nectandra sp.

Ocotea floribunda

Persea americana

Gustavia augusta

LAURACEAE - *Cinnamonum verum* J. Presl.
Canela (Cinammon)
Tree, Amazonian, 0-500m, introduced and cultivated
Use: 1. Good Luck, Love, Enchantment / Bark, dried / Topical / Mix 1 liter of water with 10g of Canela, Rose Petals (Red, White, Yellow), Ramillete de Novia, Agua Florida, Sugar and Lime Juice from 1 Lime. Boil for 2-5 minutes. Take a bath in the solution 3-4 times a month. Alternatively, grind and pulverize 100g. Rub powder over entire body while praying and wishing for the person you are yearning for, 4 times a week or as needed. **2.** Bronchitis, Recovery from all-night rituals / Bark, dried / Oral / Boil 1 garlic clove, 10g total of Matico, Veronica, Brochamelia, and Vira Vira along with 3g of Cinnamon in 1 liter of water for 3-4 minutes. Drink warm, 3-4 times a day as needed. Drink cold after a ritual in the morning during breakfast.

LAURACEAE - *Nectandra reticulata* (Ruiz & Pav.) Mez.
Ishpingo, Espingo-blanco, Espingo
Tree, Amazonian, Andean, 0-2500m
Use: 1. Nervous system / Seeds, fresh or dried / Oral / Boil 10g total of Ishpingo, Hierba de Chocho, Cordón del Muerto, Claveles, and Eucalypto in 1 liter of water for 3 min. Bathe and rub the solution over body, 3 times only. **2.** Fright/Susto / Whole plant, fresh or dried / Topical / Grind 100g of Ashango and 100g total of Ishpingo, Cedrón Seeds, Samala, and Quina Quina. Blow ground powder on patient's face, 2 times a week for 4 months.

LAURACEAE - *Nectandra* sp.
Samala, Asmala, Amala
Tree, Amazonian, Andean, 0-2500m
Use: 1. Fright/Susto, Bad Air/Mal Aire, Nervous System, Enchantment, Daño/Sorcery (Cure), Epilepsy / Seeds, dried or fresh / Topical / Grind and boil 20 Seeds in 5 liters of water for 10 mins. Bathe 2 times a week (Tuesdays and Fridays) for 1 month. **2.** Fright/Susto, Bad Air/Mal Aire, Nervous System, Enchantment, Daño/Sorcery (Cure), Epilepsy / Seeds, dried or fresh / Oral / Grind and macerate 7-15 Seeds in 1 liter of water for 8 days. Drink 3-4 small cups a day for 7 days. Seeds can also be macerated in Sugarcane Alcohol for 5 days and drunk, 5g 3 times a day for 1 week.

LAURACEAE - *Ocotea floribunda* (Sw.) Mez
Ishpino, Flor de Ishpingo, Hierba de Ishpingo
Tree, Amazonian, 0-500m
Use: 1. Bad Air/Mal Aire, Epilepsy / Seeds and Bark, fresh or dried / Oral / Mix with seeds from 6 other plants: Ashango, Pucho, Amala, Quina Quina, Nuez Mozcada, and Camalonga. Boil 5 seeds of each in 1/2 liter of water for 20 min. Drink 1 liter once a month for prevention; 7-15 days once a day for illness. **2.** Fright/Susto, Bad Air/Mal Aire, Nerves, Epilepsy, Enchantment / Leaves and Flowers, fresh or dried / Topical / Limpia: Combine 1 Leaf and 1 Flower of Ishpingo with Timolina and Vinegar. Apply 3 times a week. Bath: 20g in 3-5 liters of water boiled for 20-30 minutes and mixed with Ajenco, Ruda, Romero, Albahaca, Ortiga, Añasquero, Hierba del Aire, Hierba del Susto, Romero, and Hierba del Gallinazo (1 Leaf and 1 Flower of each), 3 times a week for illness or once a month for prevention.

LAURACEAE - *Persea americana* Mill.
Palta (Avocado)
Tree, Amazonian, Andean, 0-1500m, cultivated
Use: 1. Weight Loss / Leaves, fresh or dried / Oral / Boil 1/2 liter of water with 10 Palta Leaves for 3 minutes. Patient should drink hot solution, 1 cup 2-3 times a day for 1-2 months. **2.** Diarrhea, Kidneystones, Contraceptive, Sterilization (for women only) / Seeds, fresh / Oral / Grind 1 Palta Seed and boil in 1/2 cup of water for 3 minutes. Add Linaza and Sugar. Patient should drink lukewarm solution, 1 cup 4 times a day for a month. For a woman seeking sterility, 3 times in a row will produce the desired result. **3.** Cough / Flowers, fresh / Oral / Boil Avocado Flowers in water. Patient should drink lukewarm solution, 1/2 a cup 3 times a day for a week.

LECYTIDACEAE - *Gustavia augusta* L.
Chope
Tree, Amazonian, Andean, 0-1000m
Use: Allergies, Rashes, Pimples, Hives / Leaves, fresh / Topical / Heat 300g of Chope Leaves with 20g of Limestone in a pan for 3 min. Place warm on affected area and massage twice a day.

Lemna minuta

Linum sativum

Psittacanthus chanduyensis

Tristerix longebracteatus

Huperzia crassa

Huperzia columnaris

LEMNACEAE - *Lemna minuta* Kunth
Flor de Agua
Herb, Andean, 2000-3000m
Use: Ulcers, Stomach Inflammation. / Whole plant, fresh / Topical / Boil whole plant with water to extract 2 tbsp. Place over the stomach along with 2 Egg Whites. Use for 4 hours as a poultice.

LINACEAE - *Linum sativum* L., *Linum usitatissimum* L.
Linaza (Linseed)
Herb, Andean, Coastal, 0-3500m, cultivated
Use: Kidney Inflammation, Liver Inflammation, Prostate Inflammation, Gallbladder Stones, Kidney Stones / Seeds, dried / Oral / Mix 5g of Linaza with 10g total of Cola de Caballo, Chanca Piedra, Caña Caña, Boldo and Overo. Boil the mixture for 5 minutes. Let mixture cool. Drink 1 cup, 3-4 times a day for 2 weeks to 1 month. Drink lukewarm.

LORANTHACEAE - *Psittacanthus chanduyensis* Eichler
Suelda con Suelda
Parasitic Shrub, Coastal, 0-1000m
Use: Fractures, Twists / Leaves and Stems, fresh / Oral, Topical / Boil 5g in 1 liter of water for 3 min. Tea: 1 liter a day for 1 month. Poultice: 2 times a month.

LORANTHACEAE - *Tristerix longebracteatus* (Desr.) Barlow & Wiens
Suelda con Suelda
Parasitic shrub, Andean, 2000-4500m
Use: Bones (lacking Calcium), Vaginal Discharge, Bones (Fractured) / Whole plant, dried / Oral / Add 10g of plant material to 5g of Uña de Gato, Diego Lopez, and 1 liter of water. Boil the mixture for 4 minutes. Drink the mixture lukewarm. Take 1 cup, 3 times a day for 1 month.

LYCOPODIACEAE - *Huperzia crassa* (H. & B. ex Willd.) Rothm.
Condor, Condor Amarillo, Condorcillo, Condorcilla, Condor Rojo, Condor Verde
Clubmoss, Andean, 3000-4500m
Use: **1.** Good Luck and Success in Travels, Fragrance, Bad Air/Mal Aire / Leaves and Stems, fresh / Seguro / Place 3 small Branches with Leaves in Seguro bottle. **2.** Good Luck and Success in Travels, Fragrance, Bad Air/Mal Aire / Leaves and Stems, fresh / Topical / Boil 20g in 5 liters of water for 20 minutes combined with herbs of Strength and Luck. Bathe 2-3 times a week during the evening.

LYCOPODIACEAE - *Huperzia columnaris* B. Øllg.
Hórnamo Condor Purga
Clubmoss, Andean, 3000-4500m
Use: Laxative / Leaves and Stems, fresh / Oral / For 20 patients boil 2 San Pedros (1 of 7 ribs and 1 of 8 ribs) and 100g of Condor Purga in 4 liters of water for 3 hours. Drink cold, 1/2 cup once only.

Huperzia hohenackeri

Huperzia kuesteri

Huperzia reflexa

Huperzia sellifolia

Huperzia tetragona

Lycopodium clavatum

LYCOPODIACEAE - *Huperzia hohenackeri* (Herter) Holub
Guaminga, Huaminga
Clubmoss, Andean, 3000-4500m
Use: 1. Fright/Susto, Purgative, Bad Air/Mal Aire, Success in Work, Success in Love / Whole plant, fresh or dried / Oral / Boil 5g in 1 liter of water. Drink once a day. **2.** Fright/Susto, Purgative, Bad Air/ Mal Aire, Success in Work, Success in Love / Whole plant, fresh or dried / Seguro / Seguro: Use 7 small plants per Seguro.

LYCOPODIACEAE - *Huperzia kuesteri* (Nessel) B. Øllg.
Condor Lasio, Trensa Hermosa, Condor Crespo, Condor Cimuro, Condor
Clubmoss, Andean, 3000-4500m
Use: 1. Luck, Fragrance, Overcome Sorcery, Success in Work and Love / Whole plant, fresh or dried / Topical / Boil in 1 cup of water for 10 min 3 leaves of the following plants: Condor Cimuro, Misha Galga, and Cimora Curandera. Mix with the following 6oz perfumes: Jardín España and Tabú along with one 12oz bottle of Agua Florida and one 12oz bottle of Cananga. Drink cold. Patient must stay in a dark room, isolated and on a diet without spices for 3 days. Afterwards, patient may come out of the dark room, but must rest inside the house for another 3 days. **2.** Luck, Fragrance, Overcome Sorcery, Success in Work and Love / Whole plant, fresh or dried / Seguro / Place 7 small branches in the Seguro bottle.

LYCOPODIACEAE - *Huperzia reflexa* (Lam.) Trevis.
Condor Mulato, Enredadera
Clubmoss, Andean, 3000-4500m
Use: Fragrance, Good Luck, To prevent someone from leaving, To cause someone to return / Leaves and Roots, fresh / Topical / Boil 20g in 1 liter of water for 30 minutes with herbs for Strength. Bathe 3 times a week.

LYCOPODIACEAE - *Huperzia sellifolia* B. Ollg.
Condor Crespo
Clubmoss, Andean, 3000-4500m
Use: Luck, Fragrance / Whole plant, fresh or dried / Topical / Boil 10g in 1 liter of water for 30 min. Bathe twice a month in the evening.

LYCOPODIACEAE - *Huperzia tetragona* (Hook. & Grev.) Trevis.
Trencilla Roja
Clubmoss, Andean, 3000-4500m
Use: Fractures, Good Luck / Stems, dried / Topical / Boil 20g in 5 liters of water for 20 minutes and mix with herbs of Strength such as Hórnamos and Maiques. Bathe 3 times a week.

LYCOPODIACEAE - *Lycopodium clavatum* L.
Rastera
Clubmoss, Andean, 1500-4500m
Use: 1. Success in Work / Whole plant, fresh or dried / Oral / Boil 5g in 1 liter of water. Drink once a day. **2.** Success in Work / Whole plant, fresh or dried / Seguro / Seguro: Use 7 small plants per Seguro.

Lycopodium jussiaei

Lycopodium thyoides

Cuphea strigulosa

Punica granatum

Banisteriopsis caapi

Alcea rosea

LYCOPODIACEAE - *Lycopodium jussiaei* Desv. ex Poir
Hierba del Hombre, Rastrera
Clubmoss, Andean, 3000-4500m
Use: 1. Success in Work / Whole plant, fresh or dried / Oral / Boil 5g in 1 liter of water. Drink once a day. **2.** Success in Work / Whole plant, fresh or dried / Seguro / Seguro: Use 7 small plants per Seguro.

LYCOPODIACEAE - *Lycopodium thyoides* H. & B. ex Willd.
Trencilla Roja
Clubmoss, Andean, 3000-4500m
Use: Good Luck, Bad Air/Mal Aire, Success in Love and Business / Stems, dried / Topical / Boil 20g in 5 liters of water for 20 minutes and mix with herbs of Strength such as Hórnamos and Maiques. Bathe 3 times a week.

LYTHRACEAE - *Cuphea strigulosa* H.B.K.
Lancetilla, Sanguinaria, Hierba del Toro
Herb, Amazonian, Andean, 0-1500m, weed
Use: 1. Spiritual Flowering / Leaves and Stems, fresh / Topical / Alternative mixture for Spiritual Flowering. Bathe once. **2.** Good Luck / Leaves and Stems, fresh / Seguro / Standard Seguro mixture. **3.** Blood Circulation, Fever, Blood Purification, Intestinal Infections, Heart, Nervous System, Liver, Colic, Gas, Diarrhea, Stomach Inflammation, Kidneys, Internal Inflammation, Strengthen the Body, Anemia, Bad Air/Mal Aire / Leaves and Stems, fresh / Oral / Boil 5-20g in 1 liter of water for 3 minutes. Combine with 5g each of Congona, Claveles, Madre Selva, Ortiga, Moradilla, Contrahierba, Colores, Agujilla, Pie de Perro, Cola de Caballo, Verbena, Pimpinela, Flor Blanca, Grama Dulce, Esencia de Rosa and Cadillo. Drink 3 times a day, 1 liter daily for 1 week to 3 months. Drink solution before eating.

LYTHRACEAE - *Punica granatum* L.
Granada (Pomegranate)
Tree, Amazonian, Andean, Coastal, 0-2500m, introduced and cultivated
Use: Diarrhea / Fruit Peel, fresh / Oral / In 1 liter of water boil for 3-5 minutes 3/4 of the Fruit Peel and mix with 10g each of Hinojo and grated Palta Seed. Drink 1 glass, lukewarm 3-4 times a day for 2 weeks.

MALPIGHIACEAE - *Banisteriopsis caapi* (Spruce ex Grieseb.) Morton
Ayahuasca, Ayahuasca Verde, Ayahuasca Amarilla
Liana, Amazonian, 0-500m
Use: Enhancing Vision during rituals / Bark, fresh or dried / Oral / Boil 20g of the Bark in 1 liter of water from 12 noon until 4PM on a low fire, increasing temperature toward the end. Drink cool, 1 small cup during the ceremony. One needs to fast for 24 hours before taking the drink. Patient cannot drink during menstrual period.

MALVACEAE - *Alcea rosea* (L.) Cavanilles
Malva Blanca, Malva Morada
Herb, Andean, 2500-3500m, introduced and cultivated
Use: Inflammation, Cough, Hemorrhages / Whole plant except Stems, fresh / Oral / Boil 10g in 1 liter of water for 5 min. Drink 3 times a day as needed

Gossypium barbadense

Malva parviflora

Malva sylvestris

Theobroma cacao

Tilia platyphyllos

Urena lobata

MALVACEAE - *Gossypium barbadense* L.
Algodón Pardo, (Brown Cotton)
Herb, Amazonian, Andean, Coastal, 0-2500m, weed, cultivated
Use: 1. Evil Eye/Mal Ojo (Children) / Seed Hairs, dried / Topical / Remove the Seeds from the cotton and combine it with a mixture of the heart of 2 shredded Totora Reeds, 2 Chili Peppers and one unbroken egg. Rub the patient with the cotton and spread all over the body. Use the seedless cotton to make a sign of the cross on the patient. Then burn the cotton in a distant location. If the cotton dissolves while rubbing, the patient is very sick. Crack the egg in a glass of water and look for signs of illness in it. **2.** Wounds (External) / Seeds, fresh / Topical / Grind 200g of Seeds and extract oil. Place on top of affected area once a day until healed.

MALVACEAE - *Malva parviflora* L.
Malva Rosa, Malva Real
Herb, Amazonian, Andean, Coastal, 0-4000m, weed
Use: 1. Liver, Inflammation (General), Cough, Bronchitis, Coughing up Blood / Leaves, fresh / Oral / Combine 1 liter of water with 10g each of Pie de Perro, Chacuro, Verbena, Cola de Caballo, Amor Seco and Linaza. Also add 3-4 Leaves of Malva. Boil the mixture for 3 minutes. Patient should drink lukewarm. Drink 1 cup 3-4 times a day for 1 month. **2.** Liver, Inflammation (General), Cough, Bronchitis, Coughing up Blood / Leaves, fresh / Topical / Applied as a Poultice.

MALVACEAE - *Malva sylvestris* L.
Malva (Chica), Malva Blanca
Herb, Coastal, 0-500m, weed, introduced
Use: 1. Fright/Susto, Bad Air/Mal Aire, Heart, Nerves, Tachycardia, Epilepsy (Initial Stages) / Leaves and Stems, fresh or dried / Oral / Boil 20g in 1 liter of water for 3 minutes. Mix with 10g each of Toronjil, Pimpinela, Mejorana, Pensamiento and Cedrón. Drink 1 liter a day for 15 days. **2.** Wounds, Vaginal Cleansing / Leaves and Stems, fresh or dried / Topical / Bath: Boil 20g in 2 liters of water for 20 minutes. Wash 3 times a week. **3.** Intestinal Cleansing / Leaves and Stems, fresh or dried / Topical / Boil 10-15g each in 1 liter for 10 minutes combined with Conchalagua, Amaro, and Chicoria. Apply as Enema once a month.

MALVACEAE - *Theobroma cacao* L.
Cacao
Tree, Amazonian, Andean, Coastal, 0-1500m, cultivated
Use: Kidney Inflammation / Fruit Peel, dried / Oral / Boil 1 liter of water and add 10g of Cacao. Boil 2-3 min. Drink warm, 1 cup 3 times a day for 1 month.

MALVACEAE - *Tilia platyphyllos* Scop.
Tilo (Linden)
Tree, Andean, Coastal, 0-2500m, introduced and cultivated
Use: Nerves, Cough, Cold, Fever, Insomnia / Flowers and Leaves, fresh / Oral / Boil 1 liter of water. Add 10g each of Sauco, Manzanilla, Hinojo, Coleo, Ajenco, Toronjil, Pimpinela, and Claveles. Cover and let sit for 2-3 minutes. Patient should drink warm solution, 3-4 cups a day for 1 month.

MALVACEAE - *Urena lobata* L.
Buenas Horas
Herb, Amazonian, Andean, Coastal, 0-2000m, weed
Use: Mental Illness, Memory Loss, Confusion / Whole plant, fresh / Oral / The plant should only be gathered in the afternoon. Boil 100g of the plant in 1 cup of water for 5 minutes. Drink cold, 1/2 cup before bedtime once a day for 15 days or as needed.

Brachyotum tyrianthinum

Miconia salicifolia

Tibouchina laxa

Abuta grandifolia

Peumus boldus

Brosimum rubescens

MELASTOMATACEAE - *Brachyotum tyrianthinum* J.F. Macbr.
Sarcilleja
Shrub, Andean, 2500-3500m
Use: Blood Circulation / Stems, fresh / Oral / Boil 5g in 1 liter of water for 3-5 minutes. Drink 3 times a day for 3 days.

MELASTOMATACEAE - *Miconia salicifolia* (Bonpl. Ex Naud.) Naud.
Llatama Roja
Shrub, Andean, 2500-4500m
Use: Bad Air Mal Aire, Burns / Leaves and Stems, fresh or dried / Oral / Boil 100g in 1 cup of water for 5 minutes. Drink cold, 1/4 cup once only.

MELASTOMATACEAE - *Tibouchina laxa* (Des.) Cog.
Barbón
Shrub, Andean, 1500-3500m, weed
Use: Cataracts / Flowers, fresh / Topical / Crush and extract juice. Put in your eye as eye drops, 2 drops per eye twice a day.

MENISPERMACEAE - *Abuta grandifolia* (Mart.) Sandwith.
Abuta (Macho and Hembra)
Liana, Amazonian, 0-500m
Use: Contraceptive, Diabetes, Cholesterol / Root and Stems, fresh or dried / Oral / Boil 20-100g in 1 liter of water for 4-5 minutes. Drink warm, 1 cup 3 times a day. Drink 3 days before and 3 days after menstrual period.

MONIMIACEAE - *Peumus boldus* Molina
Boldo
Shrub, Amazonian, Andean, 0-2500m
Use: Liver Inflammation, Kidney Inflammation / Leaves, dried / Oral / Boil 1 liter of water and 10g each of Boldo, Pie de Perro, Linaza, Berros, Pata de Perro, Papa Madre, Espiga de Maíz, and Flor de Overo. Boil for 2-3 minutes. Drink warm, 1 cup 3-4 times a day for 1 month.

MORACEAE - *Brosimum rubescens* (Aublet) Huber
Palo Sangre, Palo de Sangre, Ablita
Tree, Amazonian, Andean, 0-1000m
Use: **1.** Fertility, Sexual Potency / Wood and Bark, fresh or dried / Oral / To 1 bottle of red wine add 5g each of Palo de Sangre, Palo Huaco, Cascarilla, Chuchuhuasi, Pacra, Honey, Pollen, and Huevo del Angelote. Then add 5g of Huanarpo Macho if the remedy is for a man or 5g of Huanarpo Hembra if it is for a woman. Drink 1 cup 3 times a day until bottle its finished. **2.** Blood Irrigation, Blood Coagulation, Hemorrhages (Prevention and Healing), Diabetes / Wood and Bark, fresh or dried / Oral / Chop off the small branches. Boil 5g of Ambarina in 1 liter of water along with 50g of Palo de Sangre and 50g of Palo Huaco for 10 minutes. Drink 1 liter daily for 3 months or more. **3.** Arthritis, Bronchitis, Muscle Pain / Wood and Bark, fresh or dried / Oral / Add 7 Roots or 50g to 1 bottle of Whiskey or Tequila mixed with 10g of Chuchuhuasi and Cascarilla. Drink during meals twice a day for 8-10 days.

Ficus carica

Ficus spp.

Morus alba

Muntingia calabura

Musa x *paradisiaca*

Myrica pubescens

MORACEAE - *Ficus carica* L.
Higo (Fig)
Tree, Andean, Coastal, 0-2000m, introduced, cultivated
Use: Diabetes / Leaves and Stems, fresh or dried / Oral / Boil 4 Leaves in 1 liter of water for 3 minutes. Drink lukewarm, 1 cup 3-4 times a day as needed.

MORACEAE - *Ficus* spp.
Higuerón
Tree, Amazonian, 0-500m
Use: Bones (Fractured) / Bark, fresh / Oral / Macerate in Sugarcane Alcohol. Drink 2 glasses a day.

MORACEAE - *Morus alba* L.
Morera (Mulberry)
Tree, Coastal, 0-1500m, introduced and cultivated
Use: Diabetes / Leaves and Stems, fresh or dried / Oral / Boil 4 Leaves in 1 liter of water for 3 minutes. Drink lukewarm, 1 cup 3-4 times a day or as needed for life.

MUNTINGIACEAE - *Muntingia calabura* L.
Cerezo Cimarrón
Tree, Amazonian, Andean, Coastal, 0-1500m, introduced and cultivated
Use: Gastritis, Internal Infections (General) / Fruit, fresh / Oral / Liquefy/blend 200g of the Fruit with 1/2 cup of water. Drink cold, 1 glass once a day for 6 days.

MUSACEAE - *Musa* **x** *paradisiaca* L.
Plátano (Banana)
Tree, Amazonian, Andean, Coastal, 0-2000m, Introduced and cultivated
Use: 1. Diabetes / Flowers, fresh / Oral / Boil 1 liter of water. Add 10g of Platano. Cover and boil briefly. Remove and let mixture sit for 3 minutes. Drink 1 cup, 3 times a day as needed. **2.** Asthma, Pulmonary Disease, Malaria, Dengue / Latex of the Stems, fresh / Oral / Combine 5oz of plant material, 5oz of Port Wine, 2oz of Polen, and 2 tbsp of Honey. Drink the syrup. Take 5g every 6 hours for 1 month. **3.** Wounds, Stop Bleeding / Latex of the Stems, fresh / Topical / Cut the Platano trunk with a machete. Extract Juice and collect in a container. Place extract over entire affected area. Apply once a day until the Wound is healed.

MYRICACEAE - *Myrica pubescens* H. & B. ex Wild.
Laurel
Tree, Amazonian, Andean, 0-4000m
Use: Fright/Susto, Sorcery/Daño / Leaves and Stems, dried / Topical / Bath: Boil 5g each of Laurel, Sauco, Nogal and Hierba del Susto in 3 liters of water for 10 min. Apply 2-4 times a month for washing Wounds in the morning and afternoon.

Myristica fragrans

Eucalyptus citriodora

Eucalyptus globulus

Eugenia punicifolia

Myrcianthes discolor

Psidium guajava

MYRISTICACEAE - *Myristica fragrans* L.
Nuez Moscada, Ajonjolí (Nutmeg)
Tree, Amazonian, 0-500m, Introduced and cultivated
Use: 1. Nervous System, Cough, Colic, Bad Air/Mal Aire, Asthma, Gas, Vitamins, Bronchitis, Fertility, Sexual Potency, Bone Pain / Seeds, dried / Oral / Grind Seeds and boil in 1 liter of water (1 seed to make 4 glasses). Drink 4 cups a day for 7-15 days. Alternatively, macerate 10 g of Nuez Moscada in 1 bottle of red wine with 10g each of Palo Sangre, Palo Huaco, Honey, Pacra, Huanarpo Macho, Bee Pollen, Huevo del Angelote and Para Para. Drink 1 cup in the morning, at noon, and in the evening until bottle is finished. **2.** Bad Air/Mal Aire, Epilepsy / Seeds, dried / Oral / Grind seeds and boil in 1 liter of water with seeds from 6 other plants: Ashango, Pucho, Amala, Quina Quina, Ishpingo, and Camalonga. Drink once a month.

MYRTACEAE - *Eucalyptus citriodora* Hooker
Citrodora
Tree, Andean, 2500-4000m, Introduced and cultivated
Use: Diabetes / Whole plant, fresh or dried / Oral / Boil 10g in 1 liter of water for 3 minutes. Drink 3 times a day for life.

MYRTACEAE - *Eucalyptus globulus* Labill.
Alcanfor (Camphor), Eucalipto Serrano, Eucalipto (Eucalyptus)
Tree, Andean, 2500-4000m, Introduced and cultivated
Use: 1. Bronchitis, Respiration, Cold, Cough, Sinusitis, Asthma, Rheumatism / Leaves, fresh or dried / Topical / Boil 10g each of Eucalyptus, Manzanilla, Matico, Nogal, Ajosquiro and Chilca in 1 liter of water for 10 min. Cover head and inhale for 15 min once a week for 1 month. **2.** Bronchitis, Respiration, Cold, Cough, Sinusitis, Asthma, Rheumatism / Leaves, fresh or dried / Topical / Boil 20g each of Eucalyptus, Cerraja, Borraja and Vira Vira in 10 liters of water. Manzanillon, Romero, Labanda, and Ortiga (10g each) can be added as well. Sit in the steam once a week for a mild condition, twice a week for a severe condition. **3.** Cold, Bone Pain, Congestion, Fat Burner / Leaves, fresh or dried / Topical / Bath: 500g of Eucalyptus boiled with Chilca, Palo Santo, Romero, and Ajos Giro, twice a month. Do not use too much because plant is very hot. Patient must be naked and covered with a sheet over his head, sitting to absorb the vapor for 20 minutes. Stay inside home for 24 hours after the bath, once every 30 days. 2 times only. **4.** Bronchitis, Respiration, Cold, Cough, Sinusitis, Asthma, Rheumatism / Leaves, fresh or dried / Oral / Boil 10g each of Eucaliptus, Muñaca, Escorcionera, Veronica, Humanripa, Zarzamora, and Matico in 1 liter of water for 2 min. Drink hot, 3 cups a day, for 2 weeks.

MYRTACEAE - *Eugenia punicifolia* (Kunth) DC.
Unquia Real, Rumilanchi
Tree, Andean, 500-1500m
Use: Inflammation (General) / Leaves and Stems, fresh or dried / Oral / Chop up the plant and place in 1 liter of boiling water for 2-3 minutes. Drink 3 times a day for up to 1 month.

MYRTACEAE - *Myrcianthes discolor* (Kunth) Vaughn; *Myrcianthes fragrans* (Sw) McVaugh
Lanche, Mirto
Tree, Andean, 1500-3000m
Use: Food, Memory, Brain, Vitamins (for Brain and Colds), Inflammation, Rheumatic Pain, Stomach, Menstrual Regulation / Whole plant, fresh / Oral / Boil 5g in 1 liter of water for 10 min to create a jelly or tea. Drink 1 cup at breakfast, lunch, and dinner for 1 month.

MYRTACEAE - *Psidium guajava* L.
Hoja de Guanabana (Soursop), Graviola (Guava)
Tree, Coastal, 0-2000m, introduced and cultivated
Use: Cancer, Liver / Leaves and Stems, fresh or dried / Oral / Boil 5 Leaves in 1 liter of water for 3 minutes. Drink before and after meals, 3 cups a day for 1 month.

Syzygium aromaticum

Syzygium jambos

Boerhavia coccinea

Mirabilis jalapa

Heisteria acuminata

Olea europaea

MYRTACEAE - *Syzygium aromaticum* (L.) Merr. & Perry
Clavo de Olor (Clove)
Tree, Amazonian, 0-500m, introduced and cultivated
Use: 1. Toothache / Flower Bud (clove), dried / Topical / Chew the clove near the aching tooth. Take 1-2 cloves a day as needed. **2.** Stomachache / Flower Bud (clove), dried / Oral / Boil 1/2 liter of water. Add 10 cloves. Cover and let sit for 2-3 min. Drink the infusion. Exceeding dosage can lead to kidney damage. Take mixture 2 times a day for 2-3 days.

MYRTACEAE - *Syzygium jambos* (L.) Alston
Poma Rosa
Tree, Amazonian, 0-500m, introduced and cultivated
Use: Diarrhea / Fruits and Leaves, fresh / Oral / Boil 1 cup of water and 20g of the Leaf and Fruit for 5 min. Drink cold, 1/4 cup once a day for 8 days.

NYCTAGINACEAE - *Boerhavia coccinea* Mill.
Pega Pega
Herb, Andean, Coastal, 0-2000m, weed
Use: 1. Spiritual Flowering / Whole plant, fresh / Topical / Alternative mixture for Spiritual Flowering. Take 3 baths a month in the evening. **2.** Good Business, Protection, Good Fortune, Good Health / Whole plant, fresh / Seguro / Standard Seguro mixture.

NYCTAGINACEAE - *Mirabilis jalapa* L.
Buenas Tardes
Herb, Amazonian, Andean, 0-1000m
Use: 1. Bruises, Varicose Veins / Flowers and Leaves, fresh / Topical / Place ground Leaf and Flowers on affected area and put a piece of cloth over it twice a day as needed. **2.** Kidney Disease, Inflammation (Internal), Prostate, Kidney Stones, Prostate Cancer / Root, fresh / Oral / Boil 50g of each of the following: Buenas Tardes and Paja de Lagartija (Flor de Arena) in 1 cup of water for 5 min. Add Honey. Drink lukewarm, 1 cup 3-4 times a day for 1 month.

OLACACEAE - *Heisteria acuminata* (Humb. & Bonpl.) Engler
Chuchuhuasi
Tree, Amazonian, Andean, 0-2500m
Use: 1. Cold, Cough, Bones, Arthritis / Bark, fresh or dried / Oral / Grind Bark and put in 1 bottle of red wine to macerate. Drink 1 cup 3 times a day for 15 days. Stop for 15 days. Start treatment again for 15 more days. **2.** Fertility, Sexual Potency / Bark, fresh or dried / Oral / Mix 10g of ground Bark in 1 bottle of red wine. Add Honey, and 10g each of Pacra, Huevo del Angelote, Cholitos, and Huanarpo (Macho and Hembra). Drink 1 small cup 3 times a day as needed. **3.** Arthritis, Muscle Pain, Bone Pains, Sprains, Colds, Fat Burner, Cholesterol Burner / Bark, fresh or dried / Topical / Boil 100g of Chuchuhuasi and 30g each of Eucalyptus, Molle, and Bichayo in 5 liters of water for 30 minutes. Patient must be in a closed room without clothes with a towel over the head. Patient should inhale the steam coming off the mixture and rub the body with the herbs once only or, if needed, every 3 months.

OLEACEAE - *Olea europaea* L.
Hojas de Olivo, Olivo (Olive)
Tree, Coastal, 0-500m, introduced and cultivated
Use: 1. Diabetes, Colic / Leaves, fresh / Oral / Boil 3g of Olive Leaf in 1 liter of water mixed with 3g each of Muña and Corpus Way. Drink 3 times a day for 8 days **2.** Dispelling Negative Energy from the house / Leaves, dried / Incense / Place a handful of Olivo, Mirra, Palo Santo, Incense (Copal) and Romero on top of burning charcoal to geneerate smoke for rituals. Pass the smoke over the patient's body and around the house. Repeat 2-3 times a week or as needed.

Epilobium denticulatum

Fuchsia ayavacensis

Oenothera rosea

Aa paleacea

Epidendrum calanthum

Lycaste gigantea

ONAGRACEAE - *Epilobium denticulatum* Ruiz & Pav.
Hierba Rabia
Herb, Andean, 2000-4000m
Use: Moodiness, Grumpiness, Intoxication of the Blood, Anger, Rashes from Intoxication, Ingestion of Toxic Medicine / Whole plant, fresh / Oral / Boil 5g in 1 liter of water. May combine with 5g each of Pimpinela, Cadillo, Colores, and Lancetilla. Drink 1 liter daily for 3 months.

ONAGRACEAE - *Fuchsia ayavacensis* H.B.K.
Conchalay, Conchalay Colorado
Shrub, Andean, 1500-3500m
Use: 1. Cold, Daño/Sorcery, Fright/Susto / Leaves and Stems, fresh or dried / Topical / Boil 5g each of Conchalay mixed with Sauco, Nogal, Salvia, Añasquero Grande and 7 Espiritus in 3 liters of water for 1 hour. Let cool. Take 2 lukewarm baths a week in agreement with what the Mesa indicates or twice a month. **2.** Swelling, Arthritis (Early Stages) / Topical / Leaves and Stems, fresh or dried/ Use fresh 5g fresh Leaves, combined with 5g each Conchalay Blanco, Guaminga, 7 Espiritus, Timolina, and Vinegar. Use as a Poultice twice a week in agreement with what the Mesa indicates.

ONAGRACEAE - *Oenothera rosea* Aiton
Hierba del Dominio
Herb, Andean, 1500-4000m, weed
Use: Decreasing Bad Character / Whole plant, fresh or dried / Oral / Boil 5g in 1 liter of water for 10 min. Drink 4 cups a week for 2 weeks.

ORCHIDACEAE - *Aa paleacea* (Kunth) Rchb.f.
Hierba de la Soledad, Hierba Sola
Herb, Andean, 3000-4500m
Use: 1. Depression, Loneliness / Leaves, fresh / Oral / Boil 1 Leaf in 1 cup of water. Drink once a year. **2.** Depression, Loneliness / Leaves, fresh / Seguro / One Leaf per Seguro. **3.** Contraceptive, Sterilization of Women / Leaves, fresh / Oral / Boil 3-5g of Hierba de la Soledad, in 1 liter of water mixed with 5g each Tapa Tapa and Sicana in 1 liter of water for 10 min. Drink 1 liter daily 1 week each month.

ORCHIDACEAE - *Epidendrum calanthum* Rchb. f.
Címora Negra, Címora Curandera
Herb, Amazonian, Andean, 0-2000m, weed
Use: Bad Air/Mal Aire, Fright/Susto / Leaves and Stems, dried / Oral / Boil 50g of plant material in 1 cup of water for 10 min. Drink cold once a day.

ORCHIDACEAE - *Lycaste gigantea* Lindl.
Caña Caña
Herb, Andean, 1500-2500m
Use: Kidney Inflammation / Stems, fresh / Oral / Add 10g of the plant material and 10g each of Linaza, Berro, Pata de Perro, Papa Madre, Espiga de Maiz to 1/2 liter of water. Boil the mixture for 5 minutes. Drink cold, 1/2 cup twice a day for 8 days.

Pachyphyllum crystallinum

Stelis flexuosa

Stelis sp.

Escobedia grandiflora

Oxalis bulbigera

Oxalis tuberosa

ORCHIDACEAE - *Pachyphyllum crystallinum* Lindley
Guaimi Guaimi, Huaime Huaime
Herb, Andean, 2500-4000m
Use: **1.** Self Defense. Protection / Leaves, fresh / Topical / Boil 10g in 5 liters of water for 10 min with other Strong Herbs. Apply 3 times a week. **2.** Self Defense. Protection / Stems, fresh / Seguro / Add a small Stem to the Seguro together with Herbs of Luck.

ORCHIDACEAE - *Stelis flexuosa* Lindley
Hierba del Oro, Botón de Oro
Herb, Andean, 2500-4000m
Use: **1.** Fragrance, Good Luck, Nerves, Luck in Love, Luck in Business, Good Luck for Work, Safe Travel / Whole plant, fresh / Seguro / Combine 1 Stem with Hierba de la Plata, Hierba de la Justicia, Hierba del Dominio, Encanto, Sigueme Sigueme and Herbs of Strength and Luck. **2.** Fragrance, Good Luck, Nerves, Luck in Love, Luck in Business, Good Luck for Work, Safe Travel / Whole plant, fresh / Topical / Bath: Boil 200g in 8 liters of water for 20 min. Apply 3 times a week for 1-6 months.

ORCHIDACEAE - *Stelis* sp.
Huaime-Huaime, Cucharilla
Herb, Andean, 3000-4000m
Use: **1.** Bad Air/Mal Aire, Facial paralysis caused by Mal Aire. / Whole plant, fresh / Topical / Crush the plant and heat with Agua Florida. Use 10g of the crushed plant and 1 oz of Agua Florida. Place Poultice on the opposite sides of the affected area and cover with a piece of cloth. Apply 2 times a day for 2 days. **2.** Ovarian Inflammation, Uterine Inflammation / Root, dried / Topical / Bath: Boil 5g in 1 liter of water for 3 baths a month. **3.** Good Business, Protection, Good Fortune, Good Health / Leaves and Stems, fresh / Topical / Flowering/Florecimiento: Alternative mixture for Spiritual Flowering. **4.** Good Business, Protection, Good Fortune, Good Health / Leaves and Stems, fresh / Seguro / Standard Seguro mixture.

OROBANCHACEAE - *Escobedia grandiflora* (L.f.) Kuntze
Azafrán
Herb, 1500-3000m, weed
Use: Bronchitis, Pneumonia, Chills (General) / Flowers, dried / Oral / Boil 1/2 liter of water for 3 min. with 20g of Azafrán. Drink hot, 1 cup in the morning, 1 cup in the evening for a week.

OXALIDACEAE - *Oxalis bulbigera* Knuth.
Trébol
Herb, Coastal, 0-500m
Use: Heart, Nerves, Insomnia / Leaves and Stems, fresh / Oral / Boil water. Add 10g each of Trébol, Toronjil, Poleo, Manzanilla, Hinojo and Romero. Let mixture sit for 2-3 minutes. Drink lukewarm, 1 glass 3 times a day for 1 month.

OXALIDACEAE - *Oxalis tuberosa* Molina
Oca Rosada
Herb, Andean, 2500-4500m, weed, cultivated
Use: Food, Sexual Potency / Tuber, fresh / Oral /Boil 7-8 tubers in 1 liter for 2 minutes. Drink 3 times a day for 2 weeks to a month.

Argemone mexicana

Malesherbia ardens

Passiflora caerulea

Passiflora edulis

Passiflora ligularis

Passiflora punctata

PAPAVERACEAE - *Argemone mexicana* L.
Cardo Santo
Herb, Coastal, 0-1000m, weed
Use: 1. Stomachache, Inflammation (General) / Flowers, Leaves and Stems, fresh / Oral / Boil 1 liter of water and add 10g of Cardo Santo mixed with 10g each of Cola de Caballo, Malva, Llantén, and Pie de Perro. Drink 1 cup 2-4 times a day for 1 month. **2.** Seguro de Casa, Seguro de Chacra (Farm) / Flowers, Leaves, and Stems, fresh / Amulet / Plant Cardo Santo near the house or farm and recite a prayer to ensure that the plant guards your property.

PASSIFLORACEAE - *Malesherbia ardens* J.F. Macbr.
Veronica
Shrub, Andean, 1000-2500m
Use: Cold, Cough, Bronchitis, Asthma / Whole plant, fresh or dried / Oral / Boil 5g in 1 liter of water for 10 min. combined with 5g each of Arabisca and Huamanripa (5g each). Drink 3 times a day, total 1 liter daily.

PASSIFLORACEAE - *Passiflora caerulea* L.
Pasionara
Vine, Coastal, 0-500m, introduced and cultivated
Use: Nerves, Insomnia, Anxiety / Flowers, Leaves and Stems, fresh / Oral / Boil 1 liter of water. Add 10g each of Pasionaria, Toronjil, Pimpinella, Chancas de Comida, Romero and Membrillo Peel. Let solution sit for 2-3 minutes. Drink lukewarm, 1 cup 3 times a day for 1 month.

PASSIFLORACEAE - *Passiflora edulis* Sims.
Maracuya
Vine, Amazonian, Andean, 0-1500m, introduced and cultivated
Use: High Blood Pressure / Flowers and Fruit, fresh / Oral / Add water to the Fruit. Consume hot or cold, 1 Fruit a day as needed.

PASSIFLORACEAE - *Passiflora ligularis* Juss.
Hoja de Granadilla, Granadilla
Vine, Andean, Coastal, 0-2500m
Use: 1. Liver, Blood Circulation, Inflammation (External), Kidney Inflammation, Liver Inflammation / Leaves and new Shoots, fresh / Oral / Boil 10g each of Granadilla, Boldo, Cola de Caballo, Chacur, and Amor Seco in 1 liter of water for 3-5 minutes. Drink 1 cup 3-5 times a day for 1 month. Do not use if pregnant! **2.** Diarrhea / Fruit Peel, fresh / Oral / Boil 1 liter of water. Add 3/4 of the Fruit Peel along with 10g each of Culén, Hinojo, and Chancas de Comida. Drink hot, 1 cup 3-4 times a day for 3 days or as needed.

PASSIFLORACEAE - *Passiflora punctata* L.
Tumbillo
Vine, Andean, Coastal, 0-1500m
Use: Digestion / Fruit, fresh / Oral / Eat 50g 3 times a day for 5 days.

Passiflora quadrangularis

Passiflora sp.

Gallesia integrifolia

Petiveria alliacea

Phytolacca bogotensis

Pinus patula

PASSIFLORACEAE - *Passiflora quadrangularis* L
Hojas de Tumbo
Vine, Andean, Coastal, 0-3000m
Use: Liver, Menstrual Pain, Stomach / Leaves, fresh / Oral / Add 3 Leaves to 1 liter of boiling water. Drink 3 times a day.

PASSIFLORACEAE - *Passiflora* sp.
Chulgán
Vine, Coastal, 0-1500m
Use: Promote Vaginal Dilation during childbirth. / Leaves and Stems, dried / Oral / Add 10g of plant material to 1 cup of water. Boil the mixture for 3 minutes. Drink hot, 1 cup once only.

PHYTOLACCACEAE - *Gallesia integrifolia* (Spreng.) Harms.
Palo de Ajo
Herb, Andean, Coastal, 0-1500m, weed
Use: Bronchitis, Asthma / Stems, dried / Oral / Boil 20g of Palo de Ajo in 1/2 cup of water for 2 minutes. Drink cold, 1/8 cup a day for 8 days.

PHYTOLACCACEAE - *Petiveria alliacea* L.
Mocura, Mucura
Herb, Andean, Coastal, 0-1500m, weed
Use: 1. Spiritual Flowering / Whole plant, fresh / Topical / To 5 liters of water add 1 bundle of Mocura (10g) and Yellow, Red, and White Rose Petals. Boil for to 2-3 minutes. Filter and let sit. Add 1 tsp of Sugar, Agua Florida, and Lime Juice. Bathe when lukewarm. Pray while making the sign of the cross over the body and wash with the plants. Bathe during positive energies of the full moon, no waxing or waning. Quantity is only for 1 person. Do not bathe until the following day. Bathe Tuesdays, Fridays, and Tuesdays. **2.** Protection / Whole plant, fresh / Seguro / Place 1 Stem in flask with typical Seguro plants and herbs (Hierba de la Plata, Hierba de la Fortuna, Hierba del Hallago, etc.). Take the flask with you if it is small or keep it in the house if it is large. Only the person it was intended for should handle it.

PHYTOLACCACEAE - *Phytolacca bogotensis* H.B.K.
Laylambo, Ilambo
Herb, Andean, 2000-4000m, weed
Use: 1. Daño/Sorcery, Fright/Susto, Malaria, Dengue, Yellow Fever / Flowers, Leaves and Stems, fresh / Topical / Apply fresh Leaves as a Poultice. Use infrequently since it is extremely cold.
2. Daño/Sorcery, Fright/Susto, Malaria, Dengue, Yellow Fever / Flowers, Leaves and Stems, fresh / Topical / Boil 1 bundle of fresh Leaves in 3 liters of water mixed with 1 bundle each of Añasquero, Ajenco, Sauco, Tres Hojas, and Agua de Susto. Bathe 1-2 times a week with the warm mixture making sure to rub the patient with the Leaves. Advise the patient not to rinse after the bath. May also be used in a Limpia.

PINACEAE - *Pinus patula* Schldl. & Cham., *Pinus radiata* D. Don.
Pino (Pine)
Tree, Andean, 2500-4000m, introduced and cultivated
Use: Arthritis, Rheumatism, Bone Pain / Leaves and Stems, fresh / Oral / Boil 100g of the plant for 5 minutes in 1 cup of water. Drink lukewarm, 1/4 cup once a day for 15 days.

Peperomia fraseri

Peperomia galioides

Peperomia hartwegiana

Peperomia inaequalifolia

Peperomia quadrifolia

Piper aduncum

PIPERACEAE - *Peperomia fraseri* C. DC.
Hierba de la Plata, Dolar (Dollar)
Herb, Andean, 500-1000m
Use: 1. Fragrance, Good Luck, Love, Aphrodisiac, Business, Safe Travels, Heart, Nerves, Anxiety / Flowers and Leaves, fresh / Seguro / Place 2 small Stems in a Seguro. **2.** Fragrance, Good Luck, Love, Aphrodisiac, Business, Safe Travels, Heart, Nerves, Anxiety / Flowers and Leaves, fresh / Topical / Boil 10-50g in 8 liters of water for 30 minutes combined with 10g each of Condores, Hórnamos, Trenzas, Hierba del Oro, Hierba del Cariño and Hierba de la Estrella. Bathe 3 times a week for 6 months. **3.** Heart, Nerves, Anxiety / Flowers and Leaves, fresh / Oral / Boil 10g in 1 liter of water for 3 minutes combined with 10g each of Siempre Viva, Toronjíl, Pimpinela, Romero, Mejorana, and Pensamiento. Drink 1 liter a day for 1-30 days.

PIPERACEAE - *Peperomia galioides* H.B.K.
Congonilla
Herb, Andean, 500-4000m
Use: Nerves, Bind Boyfriend or Husband, Depression, Bad Air/Mal Aire, Heart, Nervousness, Nostalgic Anxiety, Emotional Trauma / Leaves and Stems, fresh / Oral / Boil 5g in 1 liter of water for 1-2 minutes combined with 5g each of Malva Olorosa, Siempre Viva, Contrahierba, and Toronjíl. "Charge" it on the Mesa and then serve to the patient, 1 tsp 3 times during the evening and 2 cups daily for 3-4 days.

PIPERACEAE - *Peperomia hartwegiana* Miq.
Hierba de la Plata, Hierba del Tesoro
Herb, Andean, 3000-5000m
Use: 1. Good Business, Protection, Good Fortune, Good Health / Flowers and Leaves, fresh / Seguro / Standard Seguro mixture. **2.** Spiritual Flowering / Flowers and Leaves, fresh / Topicsl / Standard mixture for Spiritual Flowering.

PIPERACEAE - *Peperomia inaequalifolia* R. & P.
Congona
Herb, Andean, 500-4000m
Use: 1. Heart, Emotional Pain, Epilepsy, Forget Trauma, Forget Problems, Forget Lovesickness, Forget Bad Relationships, Anxiety, Heart Palpitations / Whole plant, fresh / Oral / Boil 5g in 1 liter of water with Congonilla, Toronjíl, Pimpinela, Mejorana and Pensamiento. The Stems and Leaves are used predominantly. Drink 3-4 times a day for 1-2 months. **2.** Heart, Emotional Pain, Epilepsy, Forget Trauma, Forget Problems, Forget Lovesickness, Forget Bad Relationships, Anxiety, Heart Palpitations / Whole plant, fresh / Topical / Burn Leaves and Inhale smoke. **3.** Heart, Emotional Pain, Epilepsy, Forget Trauma, Forget Problems, Forget Lovesickness, Forget Bad Relationships, Anxiety, Heart Palpitations / Whole plant, fresh / Oral / Put it in food to make one forget a loved one.

PIPERACEAE - *Peperomia quadrifolia* Trel.
Piri Piri (Macho and Hembra)
Herb, Andean, 500-1500m
Use: 1. Good Luck, Aphrodisiac, Good Business, Protection, Good Fortune, Good Health / Whole plant, fresh / Topical / Boil 20-50g of the Stem and Leaves of Piri Piri for 20 minutes in 3-5 liters of water plus a bit of the following: Hierba del Buen Querer, Palmerilla, Destrencilla, Lanzetilla, Hierba del Carpintero, Pega-Pega, Siempre Viva, Hierba de la Fortuna, Hierba del Tesoro, Hierba de la Plata, Hierba del Cariño, Guaime-Guaime, Hierba de la Señorita, Hierba del Caballero and Hierba de la Justicia. After boiling, add a bottle of your favorite perfume. Rub the entire body with all the herbs. Rinse with the boiled water and air dry. Do not use soap or a towel. Bathe 3 times a week. **2.** Good Luck, Aphrodisiac, Good Business, Protection, Good Fortune, Good Health / Whole plant, fresh / Topical / Add 3 liters of water to 15g of the plant material and 10g each of Hierba de la Fortuna, El Dolar, Hierba de la Plata, Chupaflor, Hierba del Halago, Tabú, and Pétalos de Rosas: Roja, Blanca, and Roja-Amarilla. Also add Agua Florida, White Sugar, and Lime Juice. Bathe Tuesday, Friday, and the following Tuesday. Patient may repeat when needed. **3.** Good Luck, Aphrodisiac, Good Business, Protection, Good Fortune, Good Health / Whole plant, fresh / Seguro / Standard Seguro mixture.

PIPERACEAE - *Piper aduncum* L.
Yerba del Soldado, Tilonga, Matico, Mogo-Mogo
Shrub, Amazonian, Andean, 0-3000m

Piper aequale

Piper nigrum

Galvesia fruticosa

Plantago linearis

Plantago major

Plantago sericea

Use: 1. Cold, Fungus, Cough, Wounds, Bronchitis, Chills, Tuberculosis, Stop Hemorrhage / Leaves, fresh or dried / Oral / Boil 5-10 Leaves in 1 liter of water for 3-5 minutes mixed with 5g each of Salvia Real, Escorcionera, Vira-Vira, Borraja, and Asma Chilca. Drink 1 liter daily for 15 days. **2.** Immune System, Infection, Inflammation, Bronchitis / Leaves, fresh / Topical / Boil 50g in 8 liters of water for 10 minutes combined with10g each of Eucalyptus, Laurel, Verbena, and Altamisa. Bathe twice a week. Alternatively, grind and pulverize 200g of the plant material. Apply the powder on affected areas, once a day until the wound is healed.

PIPERACEAE - *Piper aequale* Vahl.
Modoquero, Mogoquero
Shrub, Amazonian, Andean, 0-2500m
Use: Liver, Hepatitis, Internal Infections / Leaves and Stems, fresh or dried / Oral / Boil 5g in 1 liter of water mixed with 5g each of Flor de Overo and Boldo. Drink 3 times a day for 1 week.

PIPERACEAE - *Piper nigrum* L.
Pimienta Negra (Black pepper)
Shrub, Amazonian, 0-500m, introduced and cultivated
Use: Bronchitis / Seeds, dried / Oral / Add 10g each of Asma Chilca, Borraja, Escorcionera, Mullaca, Vira Vira, Veronica, Cinnamon, and a Clove of Garlic. Boil for 5 minutes. Drink hot. Drink 1 cup 2 times a day for 2 weeks.

PLANTAGINACEAE - *Galvesia fruticosa* J. Gmelin
Curil, Machacha
Herb, Andean, Coastal, 0-1500m, weed
Use: 1. Arthritis, Rheumatism, Nerve Pain / Flowers, Leaves and Stems, fresh / Topical / Boil 50g in 1/2 liter of Sugarcane Alcohol. Rub 1 cup daily on affected areas for 1-6 months. **2.** Cold, Bronchitis, Asthma / Flowers, Leaves and Stems, fresh / Oral / Boil 10g each of the Flowers and Stems of Curil, Zarzamora, Matico, and Nogal in 1 liter of water for 10 min. Drink 3-4 times a day for 2 weeks.

PLANTAGINACEAE - *Plantago linearis* Kunth
Llantén, Llantén de la Costa, Llantén Serrano
Herb, Andean, 1500-4000m
Use: 1. Wound Inflammation, Wounds Cleansing / Whole Plant, fresh / Topical / Boil 1 whole plant with 10 g of Matico in 1/2 liter of water. When tepid, remove the plant and apply directly to the affected area, twice a day as needed. **2.** Liver, Kidney Inflammation, Wounds, Bladder / Whole Plant, fresh / Oral / Boil 10 grams in 1 liter of water for 10 min mixed with 10g each of Cola de Caballo, Chacur, Unquia, Grama Dulce, and Flor Blanca. Drink 4 times a day for 1 month. Can harm one's vision. Don't take more than 1 month. **3.** Cough, Bronchitis / Root, fresh / Oral / Boil 2 roots in 1 liter of water for 3 minutes and combine with 10g each of Matico, Nogal, Vira Vira and Eucalypto. Drink 4 times a day as needed.

PLANTAGINACEAE - *Plantago major* L.
Llantén
Herb, Amazonian, Andean, Coastal, 0-2500m, weed, introduced
Use: 1. Hemorrhoids, Skin Tumors (Benign), Vaginal Cleansing, Wounds. / Leaves, fresh / Topical / Boil 6 Leaves in 1 liter of water for 5 minutes and mix with 10g each of Matico, Artemisa, Salvia Real, Retania and Piedra Azul. Wash once a day for 8 days. **2.** Blood Purification, Inflammation, Liver, Kidneys, Bad Breath produced by contamination of an organ / Leaves, fresh / Oral / Boil 20-30g in 1 liter of water for 3-5 minutes and mix with 10g each of Ortiga, Berros, Lancetilla, Chanca Piedra, and Flor Blanca. Drink once a day for 3-8 days. Taking too much might harm vision. **3.** Arthritis, Pain, Sprains, Contusions, Infections / Leaves, fresh / Topical / Poultice: Apply 5 Leaves with or without water, 1-2 times a day for 2-8 days. Apply warm. **4.** Bronchitis, Cough, Colic, Infected Blow or Bruise / Seeds, fresh or dried / Oral / Boil 10g or 5g in 1 liter of water. Drink 1 cup in the morning, at noon, and in the evening before eating.

PLANTAGINACEAE - *Plantago sericea* Ruiz & Pav. var. *sericea*
Rabo de Paloma, Hierba del Susto (de Cerro)
Herb, Andean, 1500-4000m

Plantago sericea var. *lanuginosa*

Plantago sericea subsp. *sericans*

Arundo donax

Cenchrus echinatus

Cymbopogon citratus

Cynodon dactylon

Use: Fright/Susto / Leaves, fresh / Oral / Boil 2-3g in 1 liter of water for 3-5 minutes. Mix with Toronjil and Mejorana (3g each). Drink 3 times a day.

PLANTAGINACEAE - *Plantago sericea* R. & P. var. *lanuginosa* Grieseb.
Pajilla Blanca
Herb, Andean, 1500-4500m
Use: Vaginal Discharge / Whole plant, fresh or dried / Oral / Boil 10g each of Pajilla Blanca, and Anga Macha in 1 liter of water for 3-4 minutes. Drink warm, 1 cup 2-3 times a day for 1 month.

PLANTAGINACEAE - *Plantago sericea* subsp. *sericans* (Pilg.) Rahn
Paja Blanca
Herb, Andean, 1500-4500m
Use: Ovarian Pain, Ovaian Inflammation, Inflammation of the Womb / Stems, fresh or dried / Oral / Boil 3g in 1 liter of water for 3-4 min. Drink 3 times a day.

POACEAE - *Arundo donax* L.
Carrizo, Caña Hueca
Herb, Andean, 1500-3000m
Use: **1.** Hemorrhoids / Whole plant, fresh / Topical / Cut fresh plant in the place where it grows. Sit by a small creek and have another person chop the tip of the plant and capture the solution released. Place it on the affected area every morning for 1 week. **2.** Eye Scratches, Eye Opaqueness / Whole plant, fresh / Topical / Soak 5 Stems overnight in 1 glass of water. Put 1-5 drops in the affected eye once a day for 3 days or as needed.

POACEAE - *Cenchrus echinatus* L.
Cadillo, Abrojo
Herb, Andean, Coastal, 0-3000m
Use: Sharp pain in any part of the body, Inflammation (General), Skin, Intestines, Liver Disease, Gallbladder Disease, Tumors, Urinary Disease / Whole plant, fresh / Oral / Boil 100g total of Cadillo, Amor Seco, Lampazo, and Trinozo in 1/2 cup of water for 3 minutes. Drink 1/4 cup once a day for 3 days.

POACEAE - *Cymbopogon citratus* (DC.) Stapf.
Hierba Luisa, Cedrón, Maria Luisa (Lemongrass)
Herb, Amazonian, Andean, Coastal, 0-2500m, introduced and cultivated
Use: Cold, Cough, Nerves, Flu, Varicose Veins, Stomach Pain, Blood Circulation, Cancer / Leaves, Roots and Stems, fresh or dried / Oral / Boil 1 liter of water. Add 5g of Hierba Luisa. Let sit for 2-3 mins. Add a little Tequila. Stems have the most strength. Drink hot solution with meals (best at breakfast).

POACEAE - *Cynodon dactylon* (L.) Persoon
Grama Dulce
Herb, Coastal, 0-1500m, weed, introduced
Use: Ovarian Cysts, Uterine Cysts, Kidney Inflammation, Inflammation (General), Uterus, Fibroids, Uterine Prolapse / Stems, dried / Oral / Boil 10g in 1 liter of water mixed with 10g each of Cola de Caballo, Verbena, Amor Seco, Malva, Flor Blanca, Hierba de Apostema, Zarzaparrilla, and Hierba del Toro. Drink 1 liter daily for 6-12 months.

Digitaria ciliaris

Gynerium sagittatum

Hordeum vulgare

Oryza sativa

Saccharum officinarum

Triticum aestivum

PAPAVERACEAE - *Argemone mexicana* L.
Cardo Santo
Herb, Coastal, 0-1000m, weed
Use: 1. Stomachache, Inflammation (General) / Flowers, Leaves and Stems, fresh / Oral / Boil 1 liter of water and add 10g of Cardo Santo mixed with 10g each of Cola de Caballo, Malva, Llantén, and Pie de Perro. Drink 1 cup 2-4 times a day for 1 month. **2.** Seguro de Casa, Seguro de Chacra (Farm) / Flowers, Leaves, and Stems, fresh / Amulet / Plant Cardo Santo near the house or farm and recite a prayer to ensure that the plant guards your property.

PASSIFLORACEAE - *Malesherbia ardens* J.F. Macbr.
Veronica
Shrub, Andean, 1000-2500m
Use: Cold, Cough, Bronchitis, Asthma / Whole plant, fresh or dried / Oral / Boil 5g in 1 liter of water for 10 min. combined with 5g each of Arabisca and Huamanripa (5g each). Drink 3 times a day, total 1 liter daily.

PASSIFLORACEAE - *Passiflora caerulea* L.
Pasionara
Vine, Coastal, 0-500m, introduced and cultivated
Use: Nerves, Insomnia, Anxiety / Flowers, Leaves and Stems, fresh / Oral / Boil 1 liter of water. Add 10g each of Pasionaria, Toronjil, Pimpinella, Chancas de Comida, Romero and Membrillo Peel. Let solution sit for 2-3 minutes. Drink lukewarm, 1 cup 3 times a day for 1 month.

PASSIFLORACEAE - *Passiflora edulis* Sims.
Maracuya
Vine, Amazonian, Andean, 0-1500m, introduced and cultivated
Use: High Blood Pressure / Flowers and Fruit, fresh / Oral / Add water to the Fruit. Consume hot or cold, 1 Fruit a day as needed.

PASSIFLORACEAE - *Passiflora ligularis* Juss.
Hoja de Granadilla, Granadilla
Vine, Andean, Coastal, 0-2500m
Use: 1. Liver, Blood Circulation, Inflammation (External), Kidney Inflammation, Liver Inflammation / Leaves and new Shoots, fresh / Oral / Boil 10g each of Granadilla, Boldo, Cola de Caballo, Chacur, and Amor Seco in 1 liter of water for 3-5 minutes. Drink 1 cup 3-5 times a day for 1 month. Do not use if pregnant! **2.** Diarrhea / Fruit Peel, fresh / Oral / Boil 1 liter of water. Add 3/4 of the Fruit Peel along with 10g each of Culén, Hinojo, and Chancas de Comida. Drink hot, 1 cup 3-4 times a day for 3 days or as needed.

PASSIFLORACEAE - *Passiflora punctata* L.
Tumbillo
Vine, Andean, Coastal, 0-1500m
Use: Digestion / Fruit, fresh / Oral / Eat 50g 3 times a day for 5 days.

Passiflora quadrangularis

Passiflora sp.

Gallesia integrifolia

Petiveria alliacea

Phytolacca bogotensis

Pinus patula

POACEAE - *Zea mays* L.
Espiga de Maíz, Chuño de Maíz, Maíz (Corn)
Herb, Amazonian, Andean, Coastal, 0-4000m, introduced and cultivated
Use: 1. Kidneys, Inflammation (Internal) / Flowers, fresh / Oral / Boil 10g of Espiga de Maíz in 1 liter of water for 3 minutes. Drink 4 times a day or as needed. **2.** Chills, Lung Pain, Kidney Inflammation / Seeds, dried / Oral / Boil 1/2 liter of water, 1/2kg of corn, and 10g each of Sugarcane Candy for 5-10 minutes until Corn is cooked. Serve hot. Reheat if not fresh. Once eaten, stay in room. Eat twice a day for 2 days. **3.** Inflammation (General), Anger Reduction / Seeds, dried / Topical / Grind 1/2kg of corn in 5 liters of water. Let it sit overnight and bathe at 6 AM without soap once a month. **4.** Indigestion, Heartburn, Stomach Acid / Leaves, fresh / Oral / Crush 100g of the plant's Leaves and Stems and filter the juice with a piece of cloth. Drink cold while fasting. Drink during breakfast, 1 small glass once a day for 10 days.

POLEMONIACEAE - *Cantua buxifolia* Jus. ex Lam.
Candu
Herb, Andean, 2500-4000m
Use: Fright/Susto, Speech Impediment / Whole plant, fresh or dried / Topical / Boil 10g of the plant in 1 liter of water and add a mixture of 10g each of Eucalyptus, Chancas de Muerto and Flor de Chochos. Bathe 3 times a week (Tuesday, Friday, Tuesday) in the afternoon. Lightly hit child on the mouth to induce speech.

POLEMONIACEAE - *Cantua quercifolia* Jus.
Dormidera, Hierba Adormecedora, Tutapure Morado (Chico)
Herb, Andean, Coastal, 0-2500m
Use: 1. Insomnia, Sedative, Good Luck, Nerves / Leaves and Stems, fresh / Oral / Boil 3-5g in 1 liter of water for 3 min. Drink 1 cup a day in the evening. **2.** Daño/Sorcery, Cutaneous Allergy / Leaves and Stems, fresh / Topical / Boil the herb in 3 liters of water and absorb vapors over half the body. Can combine with Chingue, Huaminga, and Chuque. Bathe once a month.

POLYGALACEAE - *Monnina pterocarpa* Ruiz & Pav.
Clarín
Herb, Andean, Coastal, 0-2500m
Use: 1. Throat Infection / Flowers and Leaves, fresh / Oral / Squeeze 10-15 Flowers and Leaves to remove the juice. Drink cold, 1/4 small glass once a day for 8 days. **2.** Ear Infection / Flowers and Leaves, fresh / Topical / Use same mixture for ear drops; 5 drops in each ear twice a day (6 AM and 6 PM) for 3 days.

POLYGALACEAE - *Polygala paniculata* L.
Canchalagua
Herb, Andean, Amazonian, 0-3000m, weed
Use: Blood Circulation / Whole plant, fresh or dried / Oral / Boil 3-5g in 1 liter of water for 3 min. Drink 3 times a day for 1 week.

POLYGONACEAE - *Muehlenbeckia tamnifolia* (Kunth) Meisner
Chumbiauri, Chumbiauria
Herb, Andean, 1500-4000m
Use: 1. Arthritis, Bones, Rheumatism, Sleep Aid, Cancer / Root, fresh / Oral / Boil 4 kg Chumbiauri with 10g each of Hierba de la Postema and Hierba China in 16 liters of water for 8 hours down to 2 liters. Drink 1 small cup of tea mixed with 1 cup of Honey in the evening for 1 month. **2.** Fever / Leaves, fresh / Topical / Crush and mix with Sugarcane Alcohol. Perform a Limpia twice in 1 day only.

Polygonum hydropiperoides

Rumex crispus

Grammitis moniliformis

Polypodium crassifolium

Portulaca oleracea

Portulaca pilosa

POLYGONACEAE - *Polygonum hydropiperoides* Michaux
Pica Pica
Herb, Andean, Amazonian, 0-4000m, weed
Use: 1. Uterine Infection, Kidney Inflammation / Whole plant, fresh / Oral / Boil 20g of Acelga in 1 liter of water for 10 mins. Drink 3 times a day for 1-1/2 months. **2.** Inflammation (Internal Female Organs), Vaginal Inflammation / Whole plant, fresh / Topical / Boil whole plant in 1/2 liter of water for 10 mins. Do not mix with other plants. Elevate legs in a "V" position. Pour wash into Vagina and allow to sit for 10 mins. Go to the restroom and contract vaginal muscles to expel wash. Repeat process once more immediately.

POLYGONACEAE - *Rumex crispus* L.
Acelga, Lengua de Vaca, Hojas de Mala Hierba
Herb, Andean, 1000-4000m, weed
Use: 1. Uterine Infection, Kidney Inflammation / Whole plant, fresh / Oral / Boil 20g of Acelga in 1 liter of water for 10 mins. Drink 3 times a day for 1-1/2 months. **2.** Inflammation (Internal Female Organs), Vaginal Inflammation / Whole plant, fresh / Topical / Boil whole plant in 1/2 liter of water for 10 mins. Do not mix with other plants. Elevate legs in a "V" position. Pour wash into Vagina and allow to sit for 10 mins. Go to the restroom and contract vaginal muscles to expel wash. Repeat process once more immediately.

POLYPODIACEAE - *Grammitis moniliformis* (Lag. ex Sw.) Proctor
Trencilla Pequeña
Fern, Andean, 3000-4500m
Use: Guarantee Continued Success, Bad Luck Prevention / Whole plant, fresh / Seguro / Put 4-5 small branches in a bottle.

POLYPODIACEAE - *Polypodium crassifolium* L.
Lengua de Ciervo, Lengua de Siervo, Calaguala
Fern, Andean, 3000-3500m
Use: Kidney Inflammation, Prostate, Bladder, Inflammation (Internal), Inflammation (General), Liver, Kidneys, Ulcers / Stems, fresh / Oral / Boil 10g in 1 liter water for 10 mins. with 10g total of Pie de Perro, Amor Seco, Cola de Caballo, and Pneapple Rind. Drink 1 liter daily for 1 month. Drink lukewarm.

PORTULACACEAE - *Portulaca oleracea* L. subsp. *tuberculata* Danin & H.G. Baker
Verdolaga
Herb, Amazonian, Andean, 0-1000m, weed
Use: Liver, Kidneys, Hepatitis, Liver Inflammation, Liver Cleansing / Whole plant, fresh / Oral / Boil Boldo, Flor de Arena, and Cola de Caballo in 1 liter of water for 3 mins. All these together should add up to 10g. Drink 1 cup 3-4 times a day for 1 month. Can also be eaten as a salad with olive oil, lemon, and salt once a week.

PORTULACACEAE - *Portulaca pilosa* L.
Verdolaga
Herb, Amazonian, Andean, 0-4500m, weed
Use: Hair loss, Hygiene / Root and Stems, fresh / Topical / Crush Stems and root and drain the extract to use as shampoo while showering or bathing.

Oreocallis grandiflora

Adiantum concinnum

Cheilanthes myriophylla

Jamesonia goudotii

Jamesonia rotundifolia

Pellaea ternifolia

PROTEACEAE - *Oreocallis grandiflora* (Lam.) R.Br.
Rumilanche, Huaminga
Shrub, Andean, 1500-4000m, weed
Use: 1. Ovarian Inflammation, Uterine Inflammation, Kidney Inflammation, Liver Inflammation, Arthritis, Blood / Leaves and Stems, fresh or dried / Oral / Boil 5g in 1 liter of water for 10 mins. mixed with 5g each of Flor Blanca and Flor de Arena. Drink 4 cups a day for 1 month. **2.** Sorcery by Beverage/ Daño por Brebaje, Fright/Susto / Leaves and Stems, fresh or dried / Topical / Place 1 handful in 3 liters of water. Can combine with Chingue, Poleo, Zanahoria Gentil, Conchalay Blanco, Apostema, and Trebol. Boil for 10 mins. Bathe once a week for 1 month.

PTERIDACEAE - *Adiantum concinnum* Humb. & Bonpl. ex Willd.
Culantrillo del Pozo, Culantrillo
Fern, Amazonian, Andean, Coastal, 500-3000m
Use: Blood purification, Hair loss, Menstrual regulation / Leaves and Stems, fresh or dried / Oral / 20g in 1 liter of water for 5 min, combined with Purenrosa, Lancetilla, Moradilla, and Oregano. Drink 3 glasses a day for 3 days.

PTERIDACEAE - *Cheilanthes myriophylla* Desv.
Hierba del Dominio
Fern, Andean, 3000-4500m
Use: 1. Dominating a Difficult Person, Dominating a Person Out of Control, Control Anger and Ill Humor / Leaves and Stems, fresh or dried / Seguro / Place in a bottle 10g of Hierba del Dominio, 10g of Hierba de la Justicia, and 10g of Hierba del Olvido. Seal. The patient must carry the bottle and use with prayers. **2.** Dominating a Difficult Person, Dominating a Person Out of Control, Control Anger and Il Humor / Leaves and Stems, fresh or dried / Oral / Boil 3-5g in 1 liter of water for 10 mins. Drink 3 times a day for 1 week.

PTERIDACEAE - *Jamesonia goudotii* (Hieron) C. Chr.
Hierba del Carpintero
Fern, Andean, 3000-4500m
Use: Success, To open a door in life and keep it open / Stems, fresh / Seguro / 3 Stems per Seguro.

PTERIDACEAE - *Jamesonia rotundifolia* Fée
Botón de Oro, Trencilla Amarilla, Hierba del Oso, Bonito de Oro, Bastón del Inca
Fern, Andean, 3000-4500m
Use: 1. Good Luck, Fragrance, Strength, Adornment / Whole plant, dried / Seguro / Three small Stems for 1 Seguro. **2.** Adornment, Fragrance, Strength / Leaves, fresh / Topical / Bath: Boil 200g in 15 liters of water.

PTERIDACEAE - *Pellaea ternifolia* (Cav.) Link.
Cuti Cuti, Cuti Cuti Amarillo
Fern, Andean, 3000-4500m
Use: Diabetes, Liver / Whole plant, fresh / Oral / 5g in 1 liter of water, drink 3 times a day for 1 week.

Laccopetalum giganteum

Thalictrum decipiens

Scutia spicata

Cydonia oblonga

Fragaria vesca

Geum peruvianum

RANUNCULACEAE - *Laccopetalum giganteum* (Wedd.) Ulbrich
Huamanripa, Pacra, Flor de Guarmayra
Herb, Andean, 4000-4750m
Use: 1. Cough, Bronchitis, Asthma, Flu, Cold / Leaves, fresh or dried / Oral / Place 2 small Leaves in 1/2 liter of boiling water for 10 mins. Drink 1 liter a day for 3 months. **2.** Fertilization (women) / Leaves, fresh or dried / Oral / To 1 bottle of wine add 5-6 Leaves of Pacra, 1 Ajo (Garlic), 1 Huevo del Anjelote, 3-4 spoonfuls of Honey and Pollen (Bee), 2 Cholitos (1 Hembra and 1 Macho), 1 Huanarpo (Hembra), 1 piece of Palo Sangre, and 1 skull of a Pejesapo Fish. Let sit for 1 week. Drink 3 times a day until the bottle is finished. **3.** Rheumatism / Whole plant, fresh / Topical / Macerate 1 whole Huamanripa in alcohol with 10g of Ajo Sacha, 3-4 Ajos (Machos), and 10g each of Eucalypto and Molle. Let mixture sit for a week. Rub on affected area as needed. **4.** Epilepsy, Heart Disease, Palpitations / Whole plant, fresh / Oral / Boil 10g of Huamanripa and 10g of Congona in 1/2 cup of water for 10 minutes. Drink solution cold, 1/2 cup once a day in the morning on an empty stomach for 25 days.

RANUNCULACEAE - *Thalictrum decipiens* B. Boivin
Chontilla (Chica)
Herb, Andean, 1500-4500m
Use: Fever, Mumps (Children) / Whole plant, dried / Topical / Combine 10g with 10g each of Ajenco, Salvia Real, Lailambo, 7 Espiritus and Agua del Susto. Boil 1 liter of water for 5 mins. and let cool. Apply twice a month or as illness requires.

RHAMNACEAE - *Scutia spicata* (H. & B. ex Schultes) Weberb. var. *spicata*
Pial, Pus
Tree, Andean, Coastal, 0-3000m
Use: Keeping Evil Spirits away from the house / Stems, fresh or dried / Amulet / Arrange the Stems in a cross formation and tie with a red ribbon. Place cross behind the house.

ROSACEAE - *Cydonia oblonga* Miller
Membrillo (Quince)
Shrub, Andean, 1500-2500m, introduced and cultivated
Use: 1. Depression, Nerves, Insomnia, Heart Problems / Fruit Peel, fresh / Oral / Boil 1 liter of water. Add 1 Membrillo Fruit Peel, 10g each of Manzanilla, Toronjil, Hinojo, Madre Selva, and Romero. Drink solution warm, 1 cup 3-4 times a day for 1 month. **2.** Vomiting, Nausea / Fruit Pulp, fresh / Oral / Chop the Fruit Pulp and place in 2 cups of water. Boil for 3-5 minutes. Patient should take solution slowly with a teaspoon, 1 cup every 6 hours. Use for children and pregnant women. **3.** Depression, Nerves, Insomnia, Heart Problems / Leaves, fresh / Oral / Boil in 1 liter of water. Add 10g of Membrillo Leaf, Manzanilla, Toronjil, Pimpinella, Borraja, Pensamiento, Mejorana, and Romero. Let it sit covered for 2-3 minutes. Drink 1 cup 3-4 times a day for 1 month.

ROSACEAE - *Fragaria vesca* L.
Fresa (Strawberries)
Herb, Andean, 2000-4000m, cultivated
Use: Nerves, Insomnia, Heart Disease / Leaves, fresh / Oral / Boil 1 liter of water. Add 10g of Fresa. Mix with 10g total of Hinojo, Manzanilla, and Pimpinella. Let sit for 2-3 minutes. Drink warm, 1 cup 2-3 times a day for 1 month.

ROSACEAE - *Geum peruvianum* Focke
Valeriana
Herb, Andean, 2500-4500m, weed
Use: Nerves, Insomnia / Stems and Fruits, dried / Oral / Boil 1 cup of water and 10g of Valeriana for 3 minutes. Drink cold, 1/2 cup once a day before bedtime for 15 days or as needed.

Lachemilla nivalis

Polylepis racemosa

Prunus serotina

Prunus serotina

Rosa centifolia

Rubus robustus

ROSACEAE - *Lachemilla nivalis* Kunth
Hierba del Oso
Herb, Andean, 3000-4000m
Use: Spiritual Strength and Power / Leaves and Stems, fresh / Seguro / 3 Stems per flask.

ROSACEAE - *Polylepis racemosa* R. & P.
Quinual
Tree, Andean, 2500-4500m
Use: 1. Bloating, Post-partum Detoxifier / Leaves, fresh or dried / Oral / Boil water, then add 5g of Quinual to 2 cups of hot water. Do not mix with other plants. Administer drink to the mother, no more than 2 cups. **2.** Bad Air/Mal Aire / Leaves, fresh or dried / Topical / Boil 1 bundle of Quinual with 4-5 liters of water. Do not mix with other plants. Patient should bathe with tepid water, 2 or 3 times as needed.

ROSACEAE - *Prunus serotina* Ehrh.
Helialiso
Tree, Andean, 2000-4000m, introduced and cultivated
Use: 1. Arthritis, Fractures, Bone Pain, Sprains / Leaves, fresh or dried / Topical / Mix 10g with 10g each of Altamisa, Ajenco, Ruda, and Romero and boil for 20 minutes. Bathe 3 times a week. **2.** Arthritis, Fractures, Bone Pain, Sprains / Leaves, fresh or dried / Topical / Crush 300g and macerate in alcohol. Place moist in gauze and use as Poultice for 1 week.

ROSACEAE - *Prunus serotina* Ehrhart subsp. *capuli* (Cav.) McVough
Capuli
Tree, Amazonian, Andean, Coastal, 0-4000m, introduced and cultivated
Use: Skin Lesions / Whole plant, fresh / Topical / Boil 1 liter of water containing 20g of Capuli for 3 minutes or 10g of Capuli in 1/2 liter mixed with 10g each of Verbena and Cola de Caballo. Do Not Ingest! Wash only with the water, not with the herbs, 2-3 times a day as needed.

ROSACEAE - *Rosa centifolia* L.
Rosa de Castilla, Rosa (Rose)
Shrub, Andean, 2500-4000m, introduced and cultivated
Use: 1. Laxative / Flowers, fresh or dried / Oral / Add 10g of plant material and 1 Sen Leaf to 1/2 liter of water and boil the mixture for 1-2 minutes. Drink warm, 1 cup once only. **2.** Improvement of Health, Love Life, Finances / Flowers, fresh or dried / Topical / Should collect the plant late in the afternoon. Soak in 3 liters of water 100g of Roses of each color and 100g of Margaritas. Let sit overnight and add 1 bottle of Agua Florida (12 oz) and 1 bottle of Perfume (12 oz). Rub body with Flowers and Leaves. Rinse and air dry, twice a week for 7 days.

ROSACEAE - *Rubus robustus* C. Presl.
Zarzamora, Moyaca, Zarza, Zarza Parrilla, Mora, Cushai (Blackberry)
Shrub, Andean, 2000-4000m
Use: 1. Fright/Susto, Body Pain / Flowers and Leaves, fresh or dried / Topical / Boil 10g in 5 liters of water for 30 minutes. Bathe 3 times a week. **2.** Diabetes, Cough, Cholesterol (High), Bronchitis / Flowers and Leaves, fresh or dried / Oral / Boil 3 Flower Buds in 1 cup of water mixed with 10g of Llatama. Drink 1 liter a day for 1 month. May also be inhaled. **3.** Throat (Dry), Laryngitis / Flowers and Leaves, fresh or dried / Oral / Chew like gum Moradilla, Sanguinaria, and Hierba de la Postema. Mix with Chante and Chote. May use with almost all of the other herbs. Drink 1 liter daily for 2-3 months. **4.** Kidneystones, Kidney Inflammation, Uterine Inflammation, Arthritis / Flowers and Leaves, fresh or dried / Oral / Boil 10g of leaves in 1 liter of water for 5 mins. Drink 1 liter daily for 2-3 months.

Sanguisorba minor

Arcytophyllum filiforme

Cinchona officinalis

Coffea arabica

Gardenia augusta

Uncaria guianensis

ROSACEAE - *Sanguisorba minor* Scop.
Pimpinela, Flor de Overa
Herb, Andean, 2500-4500m, weed, introduced
Use: **1.** Heart, Nervous System, Nerves, Insomnia, Depression, Heart Pain, Lovesickness, Anxiety, Menstrual Regulation, Arthritis, Blood, Fright/Susto / Whole plant, fresh / Oral / Boil 5g in 1 liter of water mixed with 10g each of Cadillo, Hierba de la Apostema, Esencia de Rosa, Lancetilla, Toronjil, Congona, Clavela, Manzanilla and Azares. Drink 3 times a day for 6-12 months. **2.** Spiritual Flowering, Good Luck / Whole plant, fresh / Topical / Standard Seguro mixture. Spray the mixture and rub the patient's body with the liquid for Good Luck. Spray the mixture every Tuesday and Thursday as needed. **3.** Spiritual Flowering, Good Luck / Whole plant, fresh / Seguro / Standard Seguro mixture. Spray the mixture and rub the patient's body with the liquid for Good Luck. Spray the mixture every Tuesday and Thursday as needed. **4.** Spiritual Flowering, Good Luck / Whole plant, fresh / Topical / Mix 50g total of: Hierba del Lucero, Hierba del Este, Ambrocilla, Señorita, Caballero, Pega Pega, Siempre Viva, Carpintero, Waime Waime, Piri Piri (Hembra y Macho), Hierba del Buen Querer, Hierba del Oro, Hierba de la Plata, Hierba del Halago, Sigueme Sigueme, and Hierba del Negocio. Boil in 5-7 liters of water for 20 mins. Then add a bit of the following perfumes: Cariño, Dios de la Huaringa, Dios de la Felicidad, San Antonio, Macumba Pusanga, Gran Jefe, Mil Flores, Llama Plata, and Ekeko letting it cool before bathing. Bathe twice (Tuesdays and Fridays only) every 3 months.

RUBIACEAE - *Arcytophyllum filiforme* (Ruiz & Pav.) Standl.
Hierba de la Madriguera
Herb, Andean, 2500-4500m, weed
Use: To prevent Overexpenditure of Money and Expenditure Non-Essentials / Whole plant, fresh / Seguro / Seguro, 1/5 of the plant per flask.

RUBIACEAE - *Cinchona officinalis* L.
Quinua, Cascarilla
Tree, Andean, 500-3500m, weed
Use: **1.** Cough / Flowers and Leaves, dried / Oral / Boil 5g in 1 liter of water mixed with 10g each of Flor Blanca, Grama Dulce and Esencia de Rosa. Drink 1 liter daily for 2 months or more. **2.** Fertility, Sexual Potency / Bark, dried / Oral / Add to a bottle of Wine or Aguardiente (Sugarcane Alcohol) 10g of Cascarilla plus 10g each of Palo de Sangre, Palo Huaco, Pacra, Piri Piri and Huanarpo. Add Honey. Drink 1 cup 3 times a day until bottle is finished. **3.** Cancer / Root, fresh / Oral / Boil 1 Root in 3 cups of water. Wait until it evaporates, leaving 1 cup. May be combined with Chumbiauria, Zarzaparrilla, Hierba de la Postema, and Poleo de la China. Take 5g a day for 14 days. **4.** Colds, Rheumatism / Bark, dried / Oral / Boil 50g of Quinua in 1 cup of water for 10 minutes. Drink lukewarm, 1/4 cup once a day for 15 days.

RUBIACEAE - *Coffea arabica* L.
Café (Coffee)
Tree, Amazonian, Andean, 0-1500m, introduced and cultivated
Use: Pain, Alertness / Seeds, dried / Oral / Boil 1/2 liter of water and filter 2-3 oz of Café into the water. Drink whenever needed.

RUBIACEAE - *Gardenia augusta* (L.) Merr.
Jasmín, Margarita
Tree, Amazonian, 0-500m, introduced and cultivated
Use:
1. Nerves, Throat Inflammation, Voice Clearance / Leaves, Stems and Flowers, fresh / Oral / Boil 1 liter of water and 10g of the Flowers for 3-5 minutes. Leave for 2-3 minutes. Drink and gargle the solution, 2-3 times a day for 2-3 days.
2. Improvement of Health, Love, Finances / Leaves, Stems and Flowers, fresh / Topical / In 3 liters of water, soak 100g of Margarita and 100g of Rose Petals. Let soak for about a day then add 1 bottle of Agua Florida and 1 bottle of your favorite Perfume. Rub body with Flowers, rinse with Agua Florida and dry. Do not use soap or towel. Perform twice a week for 7 days. A Commercial Jasmine may be used.

RUBIACEAE - *Uncaria guianensis* (Aublet) Gmelin
Uña de Gato, *Uncaria Tomentosa*, Uña de Gato de la Selva (Cat's Claw)
Liana, Amazonian, Andean, 0-1000m, weed
Use: Bronchitis, Kidneys, Asthma, AIDS, Allergies, Rheumatic Infections, Cancer, Contraceptive, Ulcers, Prostate, Bladder, Arthritis, Bones, Blood Circulation, Hemorrhages (Internal), Wounds (Internal), Kid-

Citrus aurantium

Citrus grandis

Citrus limetta

Citrus reticulata

Citrus limon

Citrus sinensis

ney Inflammation / Leaves and Stems, fresh or dried / Oral, Topical / Grind material. Better used dried. Boil 10g in 1 liter of water for 10 minutes combined with Chanca Piedra, Linaza, Boldo, Flor de Overo, and Bolsa de Pastor. Drink 1 liter daily 3 times a day for 15 days or as needed. Drink lukewarm. Solution may also be used as a Poultice. Wash wound and apply soaked Leaves.

RUTACEAE - *Citrus aurantium* (Christmann) Swingle
Hojas de Naranja (Bitter Orange)
Tree, Amazonian, Andean, 0-1200m, introduced and cultivated
Use: Nerves, Stomach / Small Leaves and Stems, dried / Oral / Boil 5g in 1 liter of water mixed with 5g each of Bolsilla de Menta and Anís. Drink 3 times a day for 1 week.

RUTACEAE - *Citrus grandis* (L.) Osbeck
Toronja (Grapefruit)
Tree, Amazonian, Andean, 0-1200m, introduced and cultivated
Use: Cholesterol, Weight Loss, Fat Burner / Fruit, fresh / Oral / Extract Juice. Dink 1 glass in the morning and 1 glass at night when needed.

RUTACEAE - *Citrus limetta* Riso
Lima (Lime)
Tree, Amazonian, Andean, 0-1200m, introduced and cultivated
Use: 1. Nerves / Flowers, fresh / Oral / Bojl 1 liter of water. Add 10g each of Lime Flowers, Manzanilla, Hinojo, Toronjil, Romero, Borraja, Madre Selva and Violeta. Drink 1 glass 3-4 times a day for 1 month. **2.** Stomach Inflammation, Gastritis, Heart Disease, Heartburn, Stomach Refeshment / Fruit, fresh / Topical / Squeeze Juice and remove the seeds. Drink 1 glass 2 times a day for 2 days. **3.** Spiritual Flowering, Sucking out Pain, Refreshing Patient, Extacting Bitterness / Fruit, fresh / Topical / Once the limes are cut, the shaman sucks out the Lime Juice and sprays over patient, taking away their bitterness and leaving him refreshed.

RUTACEAE - *Citrus reticulata* Blanco
Mandarina (Mandarin Orange)
Tree, Amazonian, Andean, 0-1200m, introduced and cultivated

Use: Nerves / Flowers and Fruit peel, fresh / Oral / Boil 1 liter of water for 2-3 minutes. Add Mandarina Flowers and Peel and 10g each of Mejorana, Toronjil, Pimpinela, Boraja, and Manzanilla. Drink 1 cup 3 times a day for 1 month.

RUTACEAE - *Citrus limon* (L.) Burm. f.
Limón (Lemon)
Tree, Amazonian, Andean, 0-1200m, introduced and cultivated
Use: 1. Nerves, Ulcer Inflammation / Flowers, fresh / Boil 1 liter of water. Add 5g of Lemon Flowers, Manzanilla, Toronjil, Pimpinella, Violeta and Claveles. Let the mixture sit for 2-3 minutes. Drink 1 glass, 3-4 times a day for 1 month or as needed. **2.** Inflammation (General), Kidney Inflammation, Ovarian Inflammation, Stomach Inflammation, Throat Inflammation, Hair loss, Dandruff / Fruit without seeds, fresh / Prepare Lemon juice by removing the peel and seeds from 3 Lemons and placing them in a can with a bit of salt. Heat the can over a fire for several seconds until Lemons become sweet. Squeeze the Lemon Juice onto the affected area (Stomach, Kidneys or Ovaries) and cover with a piece of cloth. Apply 3-4 times day for 2-3 days as needed. **3.** Inflammation (General), Kidney Inflammation, Ovarian Inflammation, Stomach Inflammation, Throat Inflammation, Hair loss, Dandruff / Fruit and Fruit Peel, fresh / Boil 1 liter of water with 1 Lemon for 2-3 mins combined with 10g each of Cola de Caballo, Pie de Perro, Chacur, Amor Seco, and Verbena. Drink 1 cup 3-4 times a day for 1 month. Solution may also be used to gargle and to wash the hair. **4.** Good Luck / Fruit and Fruit Peel, fresh / Place 7 green Lemons in a pot with 4 liters of water. Boil the mixture until the Lemons turn yellow. Remove the water and let it cool until it becomes lukewarm. Add 5g of Sugar to the temperate water. Apply mixture as a Baño de Florecimiento and a Massage. Take 2 Lemons that were submerged in the water and pray while rubbing Lemons over patient's body. Repeat until 1 Lemon is left. With 1 lemon, pray: "Out with the negative, in with the positive for my Home, Work, Love, Studies". Finish by rubbing the final lLmon over the patient's body. Discard all of the Lemons. After the bath, rinse with Cinnamon Water.

RUTACEAE - *Citrus sinensis* (L.) Osbeck
Naranja (Orange)
Tree, Amazonian, Andean, 0-1200m, introduced and cultivated

Ruta graveolens

Populus deltoides

Salix chilensis

Serjania brachyptera

Pouteria lucuma

Illicium verum

Use: 1. Depression, Nerves, Insomnia, Anxiety / Flowers, fresh / Oral / Boil 1 liter of water with 5g of Orange Flowers. Let it sit covered for 3 minutes. Mix with Toronjil, Claveles, Manzanilla, Mejorana, Chancas de Comida and Romero. Drink lukewarm, 3 cups a day for 1 month. Drink whil fasting. **2.** Laxative, especially for children / Fruit, fresh / Oral / Boil whole peel of an orange in 1 liter water for 3 - 4 minutes. Drink lukewarm, 1 cup 3 times a day for 1 week **3.** Stomach / Fruit peel, fresh / Oral / Boil whole peel of an orange in 1 liter of water for 3-4minutes. Drink lukewarm, 1 cup 3 times a day for 1 week

RUTACEAE - *Ruta graveolens* L.
Ruda, Ruda (Macho y Hembra), Hierba del Quinde
Herb, Andean, 1500-3000m, introduced and cultivated
Use: 1. Abortion, Strong Colic, Good Luck, Fright/Susto, Heart, Menstrual Regulation, Depression, Bad Air/Mal Aire, Rheumatism, Nerves, Vomiting, Nausea / Whole plant, fresh / Oral / Boil 1 liter of water with 1 whole Ruda Hembra and Genciana, Corpus Way, Salvia, Oregano, Molle, Eucalyptus, Altamisa, Ajenco, and Culantrillo. Allow to boil until there are 3 cups of solution left. For abortion use pure Ruda. Patient should drink hot solution. Drink in the morning and evening for 2 days. Be careful when ingesting because herb is extremely hot. **2.** Fright/Susto, Success, Aphrodisiac, Overcome Envy, Successful Business, Improved Life / Whole plant, fresh / Topical / Bundle the herb together with Gallinazo, Flor de Retama, Flor de Chochos, Clavel Blanco, Manzanillon, Romero, Agua del Susto, Rosas Blancas, Rosas Amarillas, Rosas Rojas, White Sugar, Hierba de la Justicia, Hierba de la Plata, Hierba de la Fortuna, Hierba del Oro, Juice of 3 Limes, Perfume Tabú, Agua Florida, and Crystallized Rock. Rub patient 3 times a month on Tuesday, Friday, and the following Tuesday. **3.** So that all goes well, To open a door in life and keep it open, To obtain Success / Whole plant, fresh / Seguro / Place whole plant in Seguro. **4.** For babies who cry too much and cannot be tranquil. / Whole plant, fresh / Oral / Crush 20 Leaves and drain the extract. Drink extract at room temperature or mix it with a glass of water, 3-4 drops once only.

SALICACEAE - *Populus deltoides* Bartram
Alamo (Poplar)
Tree, Coastal, 0-500m, introduced and cultivated
Use: Heart, Nerves, Anxiety / Leaves, fresh or dried / Oral / Boil 1 liter of water with 10g of the material for 5 mins. Add 10g each of Manzanilla, Toronjil, Pimpinella, Hinojo, Chancas de Comida, and Cascara de Membrillo. Let mixture sit for 2-3 minutes. Drink warm, 1 cup 3-4 times a day for 1 month.

SALICACEAE - *Salix chilensis* Molina

Sauce (Willow)
Tree, Amazonian, Andean, Coastal, 0-4000m, weed
Use: 1. Hangover Fever, Malaria, Colds / Leaves, fresh / Topical / Crush Leaves for juice and apply as enema once. Do not ingest. Use only when the patient is very sick. **2.** Hangover Fever, Malaria, Colds / Leaves, fresh / Oral / Boil 10g of Sauce and 10 Fruits of Capuli in 1 liter of water for 30 minutes. Drink warm, 1/2 cup every time the patient has chills.

SAPINDACEAE - *Serjania brachyptera* Radlk.
Huarate
Liana, Amazonian, Andean, Coastal, 500-1500m
Use: Diabetes, Nerves / Stems, dried / Oral / Boil 1 liter of water. Add 10g total of Manzanilla, Toronjil, Pimpinela, Hinojo, and Huarate Stems. Let mixture sit for 2 minutes. Patient should drink lukewarm solution, 1 cup 3-4 times a day for 1 month. Also used by sorcerers to for Sorcery/Daño and to burn anything that can be traced back to them.

SAPOTACEAE - *Pouteria lucuma* (R. & P.) Kuntze.
Lucuma
Tree, Amazonian, Andean, 0-3000m, cultivated
Use: Promoting Lactation in women after giving birth / Fruit, fresh / Oral / Cut 2 Fruits into pieces and boil in 2 cups of water. Boil for 4-5 minutes. Drink warm, 1 cup twice a day for 3 days.

SCHISANDRACEAE - *Illicium verum* Hook. f.
Anis Estrella
Tree, Amazonian, 0-500m, introduced and cultivated
Use: Expel residual Feces from Stomaches of newborn babies / Seeds, dried / Oral / Boil 10-15g of plant material in 1 liter of water for 2-3 minutes. Drink warm. Serve in a baby bottle, 3-4 times a day for 1-2 weeks.

Buddleja coriacea

Capraria peruviana

Siparuna muricata

Siparuna tomentosa

Smilax febrifuga

Smilax kunthii

SCROPHULARIACEAE - *Buddleja coriacea* Remy
Flor Blanca
Tree, Andean, 3000-4500m
Use: Menstruation, Ovarian Cysts, Uterine Inflammation, Inflammation (General) / Flowers, fresh or dried / Oral / Boil 5g in 1 liter of water for 3-5 mins. Mix with 10g each of Grama Dulce and Hierba de la Postema. Drink 1 liter daily for 3-8 months.

SCROPHULARIACEAE - *Capraria peruviana* Bentham
Flor de Arenilla, Té de Indio
Herb, Amazonian, Andean, Coastal, 0-2000m, weed
Use: Urine Retention, Urinary Tract Inflammation, Colic, Kidneys, Dissolve Acids / Whole plant, fresh or dried / Oral / Boil 5g in 1 liter of water for 10 mins. Drink 3 times a day.

SIPARUNACEAE - *Siparuna muricata* (R. & P.) A. DC.
Añasquero, Hojas de Añasquero, Añasquero (Grande)
Tree, Andean, 2500-3500m
Use: Fright/Susto, Arthritis, Rheumatism, Bone Pain, Muscle Pain, Stomach Pain, Daño/Sorcery, Gas, Colic / Leaves and Stems, dried / Topical / Boil 5 liters of water containing 1og each of: Añasquero, Hierba del Susto, Ishpingo, Romero, Ruda Hembra, Ishpinguillo, Chuque, Palo Santo, and 7 Espiritus for 10 minutes. Rub body with herbs and then rinse with the water. Do not dry with towel. Also used as a Poultice, 3 times a week for 1 month.

SIPARUNACEAE - *Siparuna tomentosa* (Ruiz & Pav.) A. DC.
Rinchinchin, Chinchin
Tree, Andean, 500-3500m
Use: Causing trouble for a person, Causing break-ups in a family / Leaves and Stems, dried / Topical / Grind 100g of the plant material until it is completely pulverized. Blow the powder into the face of the person to whom you want to cause trouble and mention his/her full name, once per ritual for 3 rituals.

SMILACACEAE - *Smilax febrifuga* Kunth
Palo de la China (Blanco)
Vine, Amazonian, Andean, 0-2000m
Use: Cancer (All types) / Bark, Root and Stems, fresh / Oral / Boil 50g in 2 liter of water. Wait until it evaporates, leaving 1/2 liter. May combine with 10g each of Quinua Giro and Honey. Mix also with 10g of Hierba de la Postema. Take 5g a day in the evening.

SMILACACEAE - *Smilax kunthii* Killip & Morton
Zarzaparilla, Zarza Parilla
Vine, Andean, 1500-3000m
Use: Bad Air/Mal Aire, Heart, Kidney Inflammation, Inflammation (General) / Stems, dried / Oral / In 1 cup of water boil 20g of Zarzaparrilla plus 5g each of Congona, Chacur, Matico, and Cola de Caballo for 5 minutes. Drink 1 cup in the morning before breakfast for 1 month. Repeat if necessary.

Brugmansia arborea

Brugmansia candida

Brugmansia sanguinea

Capsicum annuum

SOLANACEAE - *Brugmansia arborea* (L.) Lagerheim
Floripondio, Misha Blanca, Misha Rastrera, Misha Colambo, Datura
Tree, Amazonian, Andean, Coastal, 0-3000m, cultivated
Use: 1. Undo Sorcery, Heal the effects of Evil / Leaves, dried / Oral / Boil 3 leaves of Misha Blanca and 10 leaves of Toromaique in 1 cup of water until water is reduced to 1/2 cup. Drink cold. Patient should be kept in a dark and quiet room and on a diet of not seafood, beans or spices for 3 days. Afterwards the patient may leave the room, but should rest indoors for 3 more days. Drink 1 small cup once a day for 3 days. **2.** Protection from Evil / Whole plant, fresh / Topical / Bath mixture for Protection from Evil. Apply 3 times a day Tuesday, Friday and Tuesday. **3.** Hallucinogen, Vision Enhancement / Flowers and Leaves, fresh / Oral / Boil 30g in 8 liters of water for 1/2 hour. Effects are similar to those of Misha Roja and Misha Amarilla but the visions are weaker. Use with San Pedro and Hórnamo, 1 cup a day. Alternatively, chew 1/4 of a leaf. Overdose is lethal. **4.** Bad Air/Mal Aire / Leaves, fresh / Oral / Plant must be gathered at 6 AM. Add 2 Leaves of the plant material, 1 Leaf of Misha Amarilla, 1 leaf of Misha Roja, 1g of Toromaique, and 1g of Toromisha to 1/2 cup of water. Boil the mixture for 5 minutes. Drink the mixture cold. Patient must rest in a dark room for 3 days while maintaining a diet without spices or seafood. Exceeding the dosage is lethal. **5.** Ulcers, Cysts, Ulcers caused by Sorcery/Daño / Leaves, dried / Topical / Grind Leaves and put powder on affected area as needed. **6.** Insomnia / Flowers, fresh / Topical / Place 4 Flowers under the pillow in the form of a cross.

SOLANACEAE - *Brugmansia candida* Persoon
Misha Amarilla, Misha Galga
Tree, Andean, Coastal, 0-3000m, cultivated
Use: 1. Bad Air/Mal Aire, Diarrhea, Hallucinogen to enhance Vision / Leaves, fresh / Oral / Add 2 leaves of the plant material, 1 leaf of Misha Blanca, 1 leaf of Misha Roja, and 1g of Toromaique to 1/2 cup of water. Boil the mixture for 5 minutes. Drink the mixture cold, 1/8 of a small glass. Patient must rest in a dark room for 3 days while maintaining a diet without spices, beans or seafood. Exceeding the dosage is lethal. **2.** Ulcers, Cysts, Wounds caused by Daño/Sorcery / Leaves, dried / Topical / Grind and pulverize the Leaves. Place the powder on affected area until healed. **3.** Protection from Evil / Whole plant, fresh / Topical / Bath in the mixture for Protection from Evil. Bathe once.

SOLANACEAE - *Brugmansia sanguinea* (R. & P.) D. Don.
Misha Roja, Misha Morada, Misha Guargan, Guar Guar Rojo,
Tree, Amazonian, Andean, Coastal, 0-3000m, cultivated
Use: 1. Bad Air/Mal Aire, Protection from Daño/Sorcery, Nervous Tension, Susto from Spirits/Susto de Espíritus, Negative Spirits / Leaves and Stems, fresh / Oral / Boil 1/2 cup of water and 50g of Misha Roja for 3 minutes. Drink cold, once only. **2.** Bad Air/Mal Aire, Protection from Daño/Sorcery, Nervous Tension, Susto from Spirits/Susto de Espíritus, Negative Spirits / Leaves and Stems, fresh / Topical / Boil 6 liters of water with 10g each of: Misha Roja, Misha Blanca, Misha Amarilla, Agua de Susto, Hierba del Gallinazo, Flor de Choclo, and Toromaique for 5 minutes. Recite a prayer. Bathe the patient in the mixture while rubbing him/her with the herbs. Afterwards, rinse the patient in water, and allow him/her to air dry. Perform 3 times a day: Tuesday, Friday, and Tuesday. **3.** To see the Other World / Flowers and Leaves, fresh / Topical / Boil 6 liters of water with 10g each of: Misha Roja, Misha Blanca, Misha Amarilla, Agua de Susto, Hierba del Gallinazo, Flor de Choclo, and Toromaique for 5 minutes. Recite a prayer. Bathe the patient in the mixture while rubbing him/her with the herbs. Afterwards, rinse the patient in water and allow him/her to air dry. Perform 3 times a day: Tuesday, Friday, Tuesday.

SOLANACEAE - *Capsicum annuum* L.
Aji Panca
Herb, Amazonian, Andean, Coastal, 0-3000m, cultivated
Use: Bad Air/Mal Aire / Fruit, fresh / Incense / Mix 1kg of Aji Panca plus 1/2kg sulfur and place on top of hot burning charcoal and let the smoke spread. At this time perform the spiritual prayers. Smoke should spread around the house, room by room. No one should be in the house except the shaman alone performing the spiritual prayers.

Capsicum rhomboideum

Cestrum auriculatum

Cestrum nocturnum

Cestrum strigilatum

Datura inoxia

Jaltomata sp.

SOLANACEAE - *Capsicum rhomboideum* (Dunal) Kunze.
Aji Colorado
Herb, Amazonian, Andean, 0-2000m, cultivated
Use: 1. Evil Eye/Mal Ojo, Protection from Envy / Whole Fruit, fresh / Amulet / Tie 3 Peppers (Green, Yellow, and Red) with red ribbon and Ruda (Female and Male) for protection against Envy. Place a bunch behind the door of the house. When no longer needed, throw into the ocean or river. Hang 1 bunch behind the door until no longer needed. If in 2 days the Peppers in the bunch turn bad, there is a negative energy in the house. **2.** Evil Eye/Mal Ojo, Protection from Envy / Whole Fruit, fresh / Topical / Wrap one Aji and one whole egg in Algodón Pardo (Brown Cotton). Rub the body with water and herbs. Use the prepared bundle to rub the patient from head to toe. Then take Aji and Algodón Pardo far from the house and burn. Crack the egg and let the contents fall in a glass with water for further diagnosis.

SOLANACEAE - *Cestrum auriculatum* L'Herit
Hierba Santa, Agrasejo
Shrub, Andean, Coastal, 0-2500m, weed
Use: 1. Wounds (Ceansing) / Leaves, fresh or dried / Topical / Boil 10g of Hierba Santa in 1 liter of water for 3 minutes. Combine with 10g of Cola de Caballo. Wash with herbs and water. Use the plants to clean the wound 3 times a day until the wound heals. **2.** Fever, To lower Blood Pressure / Leaves, fresh or dried / Topical / Children: Boil 5g in 1/2 liter of water for 2 minutes. Adults: Boil 10g in 1 liter of water for 2 minutes. Add 7 Espiritus and Sugarcane Alcohol. Do not mix with other plants. Immerse body in the leaves and bath water or apply as enema, 3-4 times a month. Limpia: To increase Blood Pressure. **3.** Relaxant, Fright/Susto, Body Pain, High Blood Pressure, Typhoid Fever, Spasm Prevention after giving birth / Leaves, fresh or dried / Topical / Boil 1 bundle (20g) of Hierba Santa in 3 liters of water. Combine with 10 g each of Quinual, Eucalypto, and Romero de Campo. Wash patient with herbs and bathwater or use for a Limpia, 3 times a month as needed. **4.** Cough, Fright/Susto, Bronchitis, Colic, High Blood Pressure (To lower), Typhoid Fever, Diabetes, Liver, Cholesterol / Leaves, fresh or dried / Oral / Boil 5g in 1 liter of water with 10 g each of Corpus Way, Carqueja and Flor de Overo. Drink 1 liter a day for 7 days. **5.** Bad Air/Mal Aire, Colds, Drive off Evil Shades / Leaves, fresh or dried / Incense / Place 200g of the herb over hot charcoal. Inhale the smoke produced, once or as needed.

SOLANACEAE - *Cestrum nocturnum* L.
Flor de Azares
Shrub, Coastal, 0-1000m, introduced and cultivated
Use: Heart / Flowers, fresh / Oral / Boil 5g in 1 liter of water. Mix with 5g each of Pimpinela and Cadillo. Drink 1 liter a day for 1 month.

SOLANACEAE - *Cestrum strigilatum* R. & P., *Cestrum undulatum* R. & P.
Santa María
Shrub, Amazonian, Andean, 500-3000m
Use: Regulate Menstrual Cycle / Flowers, Leaves and Stems, fresh or dried / Oral / Boil 1 liter of water. Add 10g of Santa María, Ruda, and Orégano. Let sit for 2-3 mins. Patient should drink hot solution, 1 cup 2 times a day for 2 days.

SOLANACEAE - *Datura inoxia* Miller
Chamico
Herb, Coastal, 0-500m, introduced
Use: Bewitching Men, Lowering Morale / Leaves, dried / Oral / Buy after 6 PM for evening rituals so that it is fresh. Dry 2 Leaves. Grind into a powder. Add 1 cup of boiling water. Let sit for 3 mins. Drink hot to lukewarm as needed. Too much could kill a person. Mix with tea, chocolate or coffee as desired.

SOLANACEAE - *Jaltomata* sp.
Hierba del Tigre
Herb, Andean, 2000-3000m
Use: 1. Bad Air/Mal Aire, Fragrance, Good Luck (Work), Daño/Sorcery (Prevention), Undo bad things done to patient, Strength, Evil (Cure), Evil Eye/Mal Ojo / Leaves and Stems, fresh or dried / Oral / Boil

Juanulloa ochracea

Lycopersicon esculentum

Lycopersicon hirsutum

Lycopersicon peruvianum

Nicotiana tabacum

Solanum americanum

10g of Hierba del Tigre, 10g of Hierba del Oso, 10g of Cimora Negra, 3 Leaves of Toro Cimuro, and 3 Leaves of Misha Amarilla in 1/2 cup of water for 5 minutes. Very strong compound. Do not exceed the dosage. One eighth cup once only. Drink cold. Patient should stay inside the house without any light or noise for 3 days. Should also observe a diet (no spices, beans or seafood). **2.** Protection from Evil / Leaves and Stems, fresh or dried / Topical / Bathe for Protection from Evil. Once only. **3.** Bad Air/Mal Aire, Fragrance, Good Luck (Work), Daño/Sorcery (Prevention), Undo bad things done to patient, Strength, Evil (Cure), Evil Eye/Mal Ojo / Leaves and Stems, fresh or dried / Seguro / Place 2 small Branches in the Seguro.

SOLANACEAE - *Juanulloa ochracea* Cuatrecasas
Cuya Cuya
Shrub, Amazonian, Andean, 0-2000m
Use: Bad Air/Mal Aire, Body Pains / Seeds, dried / Topical / Grind 1 seed and mix with Agua Florida, Timolina and Sugarcane Alcohol. Rub as needed.

SOLANACEAE - *Lycopersicon esculentum* Mill.
Tomate (Tomato)
Herb, Andean, Coastal, 0-2000m, cultivated
Use: Prevention of Joint Deformation from Arthritis / Fruit, fresh / Oral / Squeeze Tomato Juice out of the Fruit. Drink cool, 1 glass a day for 1 month.

SOLANACEAE - *Lycopersicon hirsutum* Dunal
Ambulluco del Muerto
Herb, Andean, 1000-4000m, weed
Use: Fright of Death/Susto de la Muerte / Whole plant, fresh or dried / Topical / Mix with 10 g each of Flores del Muerto, Zanahoria, Poleo Gentil, Vinegar, 7 Espiritus and Agua del Susto. Perform a Limpia twice a week.

SOLANACEAE - *Lycopersicum peruvianum* L.
Tomate de Monte
Herb, Andean, Coastal, 0-3000m, weed
Use: Inflammation (Internal), Urinary Infections / Whole plant, fresh / Oral / Boil 100g of the plant material in 1 liter of water for 5 mins. Drink cold, 1/2 cup while fasting for 5 days.

SOLANACEAE - *Nicotiana tabacum* L.
Tabaco (Tobacco)
Herb, Amazonian, Andean, Coastal, 0-3000m, cultivated
Use: 1. Improve Vision and Insight during rituals, Raise Patient's Energy (Ritually and Physically) / Leaves, dried / Oral / Mix 1g of Tabaco with Agua Florida, Lime Juice, Ramillete de Novia, White Sugar, Agua Bendita, Sugarcane Alcohol, Agua Florida, and Agua Cananga. Let mixture sit for 2 hours. Both the shaman and the patient should inhale the mixture through the nose during rituals. **2.** Bone Fractures / Leaves, dried / Topical / Soak Leaves in Sugarcane Alcohol and heat. Mix with a small amount of Trementina (Turpentine). Spread and leave on Fracture. Place Poultice on affected area and cover with a piece of cloth, once only for 3 days.

SOLANACEAE - *Solanum americanum* Mill.
Hierba Mora, Hierba del Susto, Baja del Espanto
Herb, Amazonian, Andean, Coastal, 0-3000m, weed
Use: 1. Sinusitis, Flu, Cold, Involuntary Urination / Fruits fresh / Topical / Crush 20 Fruits to extract Juice, 2 drops per nostril **2.** Fever / Whole plant, fresh / Topical / Adult: Boil 10g in 1 liter of water for 10 mins. Children: Boil 1/2 liter of water per 10g 3 times a day until all mucus is released. **3.** Cold Sores, Mouth Blisters, Herpes / Fruits fresh / Topical / Squeeze Juice out of the Fruit. Apply Fruit Juice on top of the affected area, once a day until healed. **4.** Fright/Susto / Whole plant, fresh / Oral / Boil 1/2 cup

Solanum mammosum

Solanum melongena

Solanum tuberosum

Solanum sp.

Solanum sp.

Camellia sinensis

of water with 100g of the plant material for 5 minutes. Mix with 10 g each of Toronjil and Mejorana. Drink 1/4 cup, once only. **5.** Fright/Susto / Whole plant, fresh / Topical / Bath: Boil 5 liters of water with 100g of: Hierba del Susto, Hierba del Gallinazo, Romero, Paja del Aire, Ashango, Ishpingo, Samala, Flor del Muerto and Ruda. Boil for 10 minutes. First rub body with the Leaves. Second, rinse with the water. Do not use soap or water. Bathe once only on Tuesday or Friday.

SOLANACEAE - *Solanum mammosum* L.
Toro Misha Amarillo, Toro Misha, Toro Cimuro, Macumamuna
Herb, Amazonian, Andean, Coastal, 0-2000m, cultivated
Use: **1.** Good Luck, To Gain Weight, Bathing Livestock / Fruit, fresh / Topical / Boil whole Fruit with Agua Florida for 20 minutes. Bathe once a week for 1 month. Plant is highly toxic. **2.** Bad Air/Mal Aire / Leaves, fresh Whole plant, fresh / Oral / Boil 100g of Toro Misha Amarillo in 1 cup of water for 10 minutes. Drink 1 cup cold. **3.** Protection from Evil / Topical / Bath mixture for Protection from Evil.

SOLANACEAE - *Solanum melongena* L.
Berenjena (Eggplant)
Herb, Amazonian, Andean, Coastal, 0-2500m, cultivated
Use: Fat Burner, To Lose Weight / Whole Fruit, fresh / Oral / Blend 2 Berenjena with 1/4 pineapple. Drink 1 glass a day while fasting as needed.

SOLANACEAE - *Solanum tuberosum* L.
Chuño de Papa (Potato)
Herb, Amazonian, Andean, Coastal, 0-3000m, cultivated
Use: Childbirth Complications, Bronchitis, Respiratory Problems / Tuber, dried / Oral / Boil 1/2kg of Chuño de Papa in 1/2 liter of water. Add Chancaca (Sugarcane Candy), Angamacha, and Valeriana Estrella and boil for 10-15 minutes more or until the starch comes out. Remove from the flame. Serve hot as a pudding or a candy 3 times a day for 2 days within 10 days of the baby's birth. The preparation makes a kind of candy. Eat while it is cooling since it should be consumed freshly cooked. Take the last dose in bed so as not to go outside in the cold.

SOLANACEAE - *Solanum* sp.
Tutapure Chico
Herb, Andean, 1000-3000m
Use: Fright/Susto of Death, Haunting of a Ghost / Leaves and Stems, fresh / Topical / Boil 1 handful in 3 liters of water for 10 mins. Mix with Agua del Susto and 7 Espiritus. Can combine with Tutapure Grande, Tutapure Blanco, and Zanahoria, 5g each. Bathe twice a month or 1-2 times a week depending on severity. More often used for children.

SOLANACEAE - *Solanum* sp.
Hórnamo Cimuro
Herb, Andean, 1000-3000m
Use: Protection from Evil / Whole plant, fresh or dried / Topical / Bath mixture for Protection from Evil. Once only.

THEACEAE - *Camellia sinensis* (L.) Kuntze
Té (Tea)
Shrub, Andean, 1500-2500m, introduced and cultivated
Use: Colic, Diarrhea, Stomachache / Leaves and Stems, dried / Oral / Roast 20g of rice. Grind and mix with 20g of tea. Boil 1 cup of water for 5 minutes and add the ground mixture and the Juice of 3 Limes after boiling. Drink cold, 1/2 cup 2 times a day until the pain is gone.

Thelypteris scalaris

Daphnopsis weberbaueri

Tropaeolum minus

Typha angustifolia

Pilea microphylla

Urtica magellanica

THELYPTERIDACEAE - *Thelypteris scalaris* (Christ.) Alton
Helecho Macho
Fern, Andean, 1500-2500m
Use: Contraceptive / Whole plant, fresh or dried / Oral / Boil 10g of Helecho Macho with 10g of Pata de Gallina and 10g of Perejil in 1 liter of water for 10 mins. Take during the days when the woman is likely to become pregnant.

THYMELEACEAE - *Daphnopsis weberbaueri* Domke
Cholitos (Macho and Hembra)
Shrub, Andean, 3000-3500m
Use: Infertility / Seeds, dried / Oral / To 1 bottle of wine add 1 Hembra Seed and 1 Macho Seed of Cholitos, Pacra, Huanarpo (Hembra and Macho), Honey, Bee Pollen, Huevo de Angelote, Chuchuhuasi, Palo Sangre, Palo Huanco, and Cascarilla (10g of each). Let it sit for a week. Drink 1-2 small cups a day for 1 week. Drink at room temperate in the morning before breakfast and in the evening before bedtime.

TROPAEOLACEAE - *Tropaeolum minus* L.
Mastuerzo
Herb, Amazonian, Andean, Coastal, 0-3500m, weed
Use: 1. Sun spots / Flowers, fresh / Topical / Rub Flowers on affected area (usually the face). Make sure Flowers are not wet. Perform 3 times a day as needed. **2.** Stomach Inflammation / Whole plant, fresh / Oral / Boil 10g of Mastuerzo in 1 liter of water. Combine with 10g each of Amor Seco, Chacur, Cola de Caballo, Verbena, and Espiga de Maiz. Drink 3 cups a day for 1 month.

TYPHACEAE - *Typha angustifolia* L.
Chante (Cattail)
Herb, Coastal, 0-1500m, introduced
Use: Prostate / Stems, dried / Oral / Boil 5g in 1 liter of water for 10 mins. May combine with Agujilla and Achiote. Drink 1 liter daily.

URTICACEAE - *Pilea microphylla* (L.) Lieberman
Contra Hierba
Herb, Amazonian, Andean, 0-3500m, weed
Use: Bladder Stones, Kidney Inflammation, Prostate, Cysts / Whole plant, fresh / Oral / Boil 5g of the plant for 3-5 minutes in 1 liter of water mixed with 10 g each of Cola de Caballo, Hoja de Achiote, and Chanca Piedra. Drink 1 liter a day for 15 days.

URTICACEAE - *Urtica magellanica* A. Jussieu ex Poiret
Ortiga, Ortiga (Chica), Ortiga de Oveja, Ortiga Negra.
Herb, Andean, 1000-4500m, weed
Use: 1. Blood Purification, Happiness, Fever, Rheumatism, Arthritis, Blood Circulation, Hemorrhages, Hair Loss, Asthma, Hemorrhoids, Inflammation (General) / Whole plant, fresh or dried / Oral / Boil 10g of Ortiga Negra in 1 liter of water for 10 mins. Combine with 10g each of Huamanripa, Veronica, Corpus Way, Carqueja, Nogal, Ocalito, Molle, Ruda, and Matico. Drink 4 cups a day for 15 days. **2.** Blood Purification, Happiness, Fever, Rheumatism, Arthritis, Blood Circulation, Hemorrhages, Hair Loss, Asthma, Hemorrhoids, Inflammation (General) / Whole plant, fresh or dried / Topical / Use same mixture for bath and rub leaves on parts afflicted with Rheumatism 3 times a week.

Urtica urens

Aloysia triphylla

Lantana scabiosaeflora

Lippia alba

Verbena litoralis

VERBENACEAE

URTICACEAE - *Urtica urens* L.
Ortiga
Herb, Amazonian, Andean, Coastal, 0-4500m, weed, introduced
Use: Bad Air/Mal Aire, Prostate, Fright/Susto, Vaginal Cleansing, Business, Overcoming Bad Luck, Fright in Children/Susto en Niños / Stems and Leaves, fresh / Topical / Boil 7 Stems with their Leaves in 3-5 liters of water combined with 10g each of Agua del Susto, Ajenco, and Llatama for 20 mins. Bathe 2-3 times a week at 7, 9, and 11PM.

VERBENACEAE - *Aloysia triphylla* (L'Her.) Britt.
Cedrón, Pepas de Cedrón,
Tree, Andean, 2000-4000m, weed
Use: 1. Depression, Nerves, Insomnia, Stomach / Whole plant, fresh / Oral / Boil 1 liter of water. Add Cedrón and mix with Chancas de Comida, Toronjil, Hinojo, Madre Selva, Claveles, and Pensamiento. Use a total of 10g for all the material. Let the mixture sit for 2-3 mins. Patient may drink solution at any temperature, but lukewarm is recommended. Drink 1 cup, 3-4 times a day for 1 month. **2.** Bad Air/Mal Aire / Seeds, fresh or dried / Topical / Grind 4-5 Seeds and boil in 5 liters of water for10 mins. Bathe. **3.** Bad Air/Mal Aire / Seeds, fresh or dried / Oral / Boil 1 Seed in 1 liter of water for 10 mins. Drink 4 cups a day for 7 days. **4.** Sharp Body Pain / Seeds, fresh or dried / Oral / Mix 10g of ground Seeds, 1/4 of a small cup of Pisco, 1g of Alucena, 1g of Orégano, and 1g of Pimienta. Heat. Drink 1 tblsp a day for 6 days.

VERBENACEAE - *Lantana scabiosaeflora* Kunth
Mastrando, Mastrante
Tree, Andean, Coastal, 0-4000m
Use: Cold, Ovaries, Menstruation, Women after childbirth to avoid colds / Leaves and Stems, fresh or dried / Oral / Boil 20-100g in 1 liter of water for 3 mins. Mix 10g each of Canchalagua, Culantrillo, Purenrosa, Panisara, and Salvia Real. Drink lukewarm 1 liter a day for 3 days. This treatment is only for women.

VERBENACEAE - *Lippia alba* (Miller) N.E. Brown
Poleo del Inca
Shrub, Amazonian, 0-500m, weed, cultivated
Use: Cold, Colic, Kidney Inflammation, Bronchitis, Rheumatism, Gas / Leaves and Stems, fresh / Oral / Boil 5g in 1 liter of water for 10 mins. Drink 1 liter daily for 1 month.

VERBENACEAE - *Verbena litoralis* Kunth
Verbena
Herb, Amazonian, Andean, Coastal, 0-4000m, weed
Use: 1. Fever, Fungus / Whole plant, fresh or dried / Topical / Adults: Boil 10g of Verbena in 1 liter of water. Children: Boil 10g of Verbena in 1/2 liter of water. Apply enema when water is lukewarm once only. Alternatively, boil 10g per 2 liters of water for 30 minutes combined with Matico, Malva, Llantén, and Para Para, 3 times a day for 8 days. **2.** Hyperactivity, ADHD, Tranquility / Whole plant, fresh or dried / Topical / Boil for 20 minutes 5-10g of the herb per 1 liter of water mixed with 10g each of Matico, Malva, and Manzanilla. Bathe 3 times a week. **3.** Inflammation, Wounds (Cleansing), Blood Purification, Cholera, Modify strong character, Colic, Colds / Whole plant, fresh or dried / Oral / Boil 30g in 1 liter of water for 3 mins. Mix with 10g each of Cerraja, Moradilla, and Verdolaga. Drink 2 glasses a day for 4 days in the morning and at night.

VERBENACEAE
Llantama Blanca
Use: Foot Blisters / Leaves and Stems, dried / Oral / Boil 100g of plant material in 1 cup of water for 10 minutes. Drink cold, 1/4 cup twice a week.

Viola tricolor

Vitis vinifera

Aloe vera

Ximenia americana

Xyris subulata

Zingiber officinale

VIOLACEAE - *Viola tricolor* L.
Pensamiento Amarillo, Hierba del Pensamiento, Hierba del Tacón (Violet)
Herb, Andean, Coastal, 0-4000m, introduced and cultivated
Use: Heart, Lovesickness, Nerves, Insomnia, Forgetting Pain, Fright/Susto, Bad Air/Mal Aire, Nerves, Epilepsy / Whole plant, fresh or dried / Oral / Boil 1 liter of water. Add 10g of Pensamiento Amarillo to 10 g each of plants for the heart, including Toronjil, Congona, and Clavel. Drink 3 cups a day as needed.

VITACEAE - *Vitis vinifera* L.
Uva (Grape)
Liana, Andean, Coastal, 0-2000m, introduced and cultivated
Use: Bronchitis, Laxative / Fruits, dried / Oral / Add 1/2 liter of fresh milk to 10g of Dried Grapes (Raisins). Boil the mixture for 3-4 minutes. Drink hot. Drink 1 glass 3 times a day for 2 weeks.

XANTHORROEACEAE - *Aloe vera* (L.) Burm f.
Sábila, Zábila, Aloe, Hojas de Sábila, Aloe Vera
Herb, Amazonian, Andean, Coastal, 0-3000m, introduced and cultivated
Use: 1. Inflammation (External), Vaginal Inflammation, Vaginal Ulcers, Vaginal Cancer, Hair Growth, Skin Embellishment, Cataracts, Eyes, Wounds, Burns / Leaves, fresh / Topcial / Take a wide leaf and remove the spines. Cut the Leaf longitudinally and remove the iodine secretion. Heat and apply over inflamed area as Poultice 2 times a day for 1 week. For Vaginal Inflammation, insert Leaf 2-3 times a day or as needed. Leave it inside for 5 minutes. In case of Hair Loss apply to hair and skin. Juice can also be applied to eyes, 1 drop in each eye every 2 days for 6 days. **2.** Weight Loss, Gastritis, Inflammation, Diabetes, Cough, Bronchitis, Kidneys, Ulcers, Cholesterol, Cancer, Asthma, Bile / Leaves, fresh / Oral / Combine 1kg of herb, 1/2kg of Honey, and 3 tbsp of pisco. Open the leaf longitudinally and extract the iodine secretion and the internal gel from the inside of the leaf. Consume the iodine secretion and the gel, 1-2 cups a day for a week to a month. Leaf can also be macerated in a bottle of alcohol. **3.** Good Luck, Happiness / Leaves, fresh / Amulet / Hang whole plant over doorway.

XIMENIACEAE - *Ximenia americana* L.
Limoncillo
Tree, Amazonian, 0-1000m
Use: Nerves, Stomach, Menstrual Regulation / Whole plant, fresh or dried / Oral / Boil 1 liter of water. Add 10g total of Limoncillo, Panisara, Inojo, Ajenjo, Toronjil, and Pimpinela. Let mixture sit for 3 min. Patient may drink at any temperature. Drink 1 cup, 3-4 times a day for 1 month.

XYRIDACEAE - *Xyris subulata* Ruiz & Pav.
Hierba del Caballero, Chupaflor
Herb, Andean, 2500-4500m
Use: Wooing a Woman, Fragrance, Good Luck, Good Business, Protection, Good Fortune, Good Health / Stems, fresh / **1.** Seguro / Standard Seguro mixture / **2.** Topical / Standard mixture for Spiritual Flowering.

ZINGIBERACEAE - *Zingiber officinale* Roscoe
Kion, Gengibre (Ginger)
Herb, Amazonian, Andean, Coastal, 0-2000m, introduced and cultivated
Use: Cold, Cough, Bronchitis / Root, fresh / Oral / Cut Kion into small pieces and add 10g of this to 10 g each of Matico, Nogal, and Veronica. Boil in 1/2 liter of water. Drink 1 cup 3 times a day for 1 week.

ZINGIBERACEAE

Tribulus terrestris

Anti Ajo

Giartina chamissoi, Giartina glomerata, Giartina paitensis

Siphula sp.

ZINGIBERACEAE

Chimapampana

Use: To Sexually Excite and Trap a member of the opposite sex / fresh whole plants / Topical / Use the red or purple tuber. Grate to create a light perfume. Put 1/2 tuber in ointment container. Mix with Cariño Perfume. Place Perfume on the body as needed.

ZYGOPHYLLACEAE - *Tribulus terrestris* L.

Abrojo, Cadillo

Herb, Andean, Coastal, 0-2000m, weed

Use: Sharp Body Pain, Inflammation (General), Skin, Intestine, Liver Disease, Gallbladder Disease, Tumors, Urinary Disease / Whole plant, fresh / Oral / Boil 10g each of Abrojo, Amor Seco, Lampazo, and Trinoso in 1/2 cup of water for 3 minutes. Drink 1/4 cup once a day for 3 days.

UNIDENTIFIED

Anti Ajo

Use: Sorcery within a House / Stems, fresh / Amulet / Boil 40cm of the Anti Ajo Stems in 2 liters of water for 3 hours. Let sit for 5 minutes. Use Ruda (Hembra and Macho) and dip into solution and distribute in every corner of the house, always going from right to left. Splash the remainder of the solution in front of the house.

UNIDENTIFIED

Hierba del Hongo

Use: Concussion, Anemia, Liquid Build-Up in Tissues or Wounds, Kidneys / Whole plant, fresh or dried / Topical / Boil 10g for 5 minutes in 1 liter of water or 5g in 3 liters. Add 2 tbsp of Vinegar, Lemon Juice, Manzanilla Blanca, Laurel, and Llantén. Washing solution for feet, hands, and other body parts, 1 wash a day in the afternoon until symptoms are alleviated.

ALGAE - *Giartina chamissoi, Giartina glomerata, Giartina paitensis*

Cochayuyo, Mococho

Use: Weight Loss, Cholesterol / Whole plant, fresh / Oral / Steam 20g of the plant material in hot water. Blend the steamed material with the Juice of 3-4 Oranges. Drink 1 glass twice a day for 1 month.

LICHENES - *Siphula* sp.

Papelillo, Papelilla, Palalio

Use: Liver / Leaves, fresh / Oral / Boil 5g in 1 liter of water for 10 mins.

NON-PLANT MATERIAL

Polen de Zapote, Polen de Espina Negra, Polen de Arboles, Polen de Ciachon (Insect Feces)

Use: Bronchitis, Asthma, Tuberculosis / Insect Feces / Oral / Put 4g of the Feces of the Larvae ("Pollen") into 1 liter of boiling water. Is very strong. Drink 1 liter daily for 1 month.

Standard Seguro Mixture

Mix all of the of the following in a bottle of perfume: 3 Stems and Leaf of Hierba de la Señorita plus a bit of the following: Hierba del Buen Querer, Palmerilla, Destrencilla, Lanzetia, Hierba del Carpintero, Pega-Pega, Siempre Viva, Hierba de la Fortuna, Hierba del Tesoro, Hierba de la Plata, Hierba del Cariño, Guaime-Guaime, Piri- Piri, Hierba del Caballero, and Hierba de la Justicia. Amount depends on the size of the bottle. Add "Cariño" perfume and a bit of the following perfumes: Dios de la Huaringa, Dios de la Felicidad, San Antonio, Macumba Pusanga, Gran Jefe, Mil Flores, Llama Plata, and Ekeko. The shaman will then bless and spray it. Keep the Seguro at home or, if it is small enough, carry it with you at all times.

Standard Mixture for Spiritual Flowering

Boil 3 liters of water for 30 minutes with 10g of Chupaflor and 10g each of Hierba del Buen Querer, Palmerilla, Destrencilla, Lanzetia, Hierba del Carpintero, Pega Pega, Siempre Viva, Hierba de la Fortuna, Hierba del Tesoro, Hierba de la Plata, Hierba del Cariño, Guaime Guaime, Piri Piri, Hierba de la Señorita, Hierba de la Justicia, Hierba de la Fortuna, El Dolar, Hierba de la Plata, Chupaflor, Hierba del Halago, and Pétalos de Rosas Rojas, Blancas, and Roja-Amarillas. Also add Agua Florida, Tabú, White Sugar, and Lime Juice. After boiling, add a bottle of your favorite perfume. Bathe 3 times: Tuesday, Friday, and the following Tuesday. Patient may repeat when needed. Rub the entire body with all the herbs. Rinse with the mixture and air dry. Do not use soap or a towel.

Alternative Mixture for Spiritual Flowering

In 3 liters of water boil for 30 minutes 3 Stems and 1 Leaf of Hierba de la Señorita plus a bit of the following: Hierba del Buen Querer, Palmerilla, Destrencilla, Lanzetia, Hierba del Carpintero, Pega-Pega, Siempre Viva, Hierba de la Fortuna, Hierba del Tesoro, Hierba de la Plata, Hierba del Cariño, Guaime-Guaime, Piri-Piri, Hierba del Caballero, and Hierba de la Justicia. After boiling, add a bottle of your favorite perfume, Rub the entire body with all the herbs. Rinse with the mixture and air dry, once only. Do not use soap or a towel.

Bath Mixture for Protection from Evil

Boil 6 liters of water for 5 minutes with 10g each of: Misha Blanca, Misha Morada, Misha Roja, and Toromaique. Recite a prayer. Bathe the patient in the mixture while rubbing him or her with the herbs. Afterwards, rinse the patient in water and allow him or her to dry naturally.

Bibliography

Abad L (2003) **Etnocidio y Resistencia en la Amazonia Peruana**. Ediciones de la Universidad de Castilla-La Mancha, Cuenca-Spain.

Acosta J (1588-90) **The Natural and Moral History of the Indies**. Reprinted form the English edition of Edward Grimson (1604), vols. 1-2. Hakluyt Society, London; 1880.

Agurto J (2005) **Comentarios de Panelistas.** In: Ferro P, Ruiz M (Eds) *¿Como prevenir la biopirateria en el Peru? Reflexiones y Propuestas* pp. 70-73. Lerma Gomez E.I.R.L., Lima-Miraflores.

Alam G, Belt J (2009) **Developing a medicinal plant value chain: Lessons from an initiative to cultivate kuti *(Picrorhizakurrooa)* in Northern India.** *KIT Working Papers Series* pp. 1-14. The Royal Tropical Institute (KIT), The Netherlands.

Alarco de Zandra A (1988) **Perú, el libro de las plantas mágicas**. Concytec, Lima.

Alban C (1985) **Un Registro de Datos Etnobotánicos.** *Boletín de Lima* 7(39):93-96.

Alcedo A de (1786-89) **Diccionario geográfico histórico de las indias occidentales o América**. Madrid.

Alexiades MN (1996) **Collecting ethnobotanical data: an introduction to basic concepts and techniques.** In: Alexiades MN (Ed). *Selected Guidelines for Ethnobotanical Research: A Field Manual* pp. 53-94. The NewYork Botanical Garden, New York.

Alva W (1994) **Sipán: descubrimiento e investigación. Cultura y artes del antiguo Perú**. Backus & Johnston, Lima.

Alva W, Donnan CB (1993) **Royal Tombs of Sipán**. Fowler Museum of Cultural History, University of California, Los Angeles.

Alva W, Donnan CB (1994) **Tales from a Peruvian Crypt**. *Natural History* 103(5):26-36.

Angulo P (2009) **Nuevos enfoques en la investigación de plantas medicinales.** In: Sánchez Garrafa R & R (Eds) *Medicina Tradicional Andina: Planteamientos y aproximaciones* pp. 351-384. Centro de Estudios Regionales Andinos Bartolomé de las Casas (CBC)/Centro de Medicina Andina (CMA), Cuzco.

Baker J, Borris R, Carté B, Cordell G, Soejarto D, Cragg G, Gupta M, Iwo M, Madulid D, Tyler V (1995) **Natural Product Discovery and Development, New Perspectives on International Collaboration**. *Journal of Natural Products* 58(9):1325-1357.

Balick M, Mendelsohn R (1992) **Assessing the Economic Value of Traditional Medicines from Tropical Forests.** *Conservation Biology* 6:128-129.

Bannister K, Barrett K (2001) **Challenging the Status Quo in Ethnobotany: A New Paradigm for Publication May Protect Cultural Knowledge and Traditional Resources**. *Cultural Survival Quarterly* 24(4):10-12.

Barve V, Bhatti R, Bussmann RW, Bye R, Chen J, Dulloo E, Giovannini P, Linares E, Magill R, Roguet D, Salick J, On TV, Vandebroek I, Wightman G, Wyse Jackson P (2013) **A Global Program on Conservation of Useful Plants and Traditional Knowledge - A Call to Action.** https://www.researchgate.net/publication/236633499_A_Global_Program_on_Conservation_of_Useful_Plants_and_Traditional_Knowledge_A_Call_to_Action?ev=prf_pub

Bastien J (1987) **Healers of the Andes: Kallawaya Herbalists and Their Medicinal Plants**. University of Utah Press, Salt Lake City.

Bazán S (2005) **Propiedad Intelectual: Conceptos Básicos.** In: Ferro P, Ruiz M (Eds) *Cómo prevenir la Biopiratería? Reflexiones y Propuestas* pp 21-35. Lerma Gómez E.I.R.L, Lima-Miraflores.

Béjar E, Bussmann RW, Roa C, Sharon D (1997) **Pharmacological Search for Active Ingredients in Medicinal Plants of Latin America**. In: Shuman T, Garrett M, Wozniak L (Eds), *International Symposium on Herbal Medicine, A Holistic Approach* pp. 63-81, SDSU International Institute for Human Resources Development, San Diego.

Béjar E, Bussmann RW, Roa C, Sharon D (2001) **Herbs of Southern Ecuador – Hierbas del Sur Ecuatoriano**. Latin Herbal Press, San Diego.

Bermúdez A, Velásquez D (2002) **Etnobotánica de una Comunidad Campesina del Estado de Trujillo: Un Estudio Preliminar Usando Técnicas Cuantitativas.** *Revista de la Facultad de Farmacia* 44:2-6.

Bianchi A, Samorini G (1993) **Plants in Associacion with Ayahasuca.** *Jahrbuch Ethnomedizin* 1993:21-42.

Brack Egg A (1999) **Diccionario enciclopédico de plantas útiles del Perú**. PNUD – CBC, Cuzco.

Brack Egg A (2004) **Biodiversidad, pobreza y bionegocios**. PNUD, Lima.

Brako L, Zarucchi JL (Eds) (1993) **Catalogue of the Flowering Plants and Gymnosperms of Peru**. Missouri Botanical Garden, Saint Louis, MO.

Breevort P (1998) **The Booming U. S. Botanical Market, A New Overview.** *HerbalGram* 44.

Bristol ML (1969) **Tree Datura drugs of the Columbian Sibundoy.** *Botanical Museum Leaflets* 22:165-227.

Brown M (2003) **Who Owns Native Culture?** Harvard University Press, Cambridge.

Brüning HH (2004) **Diccionario Mochica**. Universidad de San Martin de Porres, Lima.

Brush S, Stabinsky D (Eds) (1996) **Valuing Local Knowledge: Indigenous People and Intellectual Property Rights**. Island Press, Washington, D.C.

Buitron X (1999) **Ecuador, uso y comercio de plantas medicinales, situacion actual y aspectos importantes para su conservación**. TRAFFIC International, Cambridge, UK.

Burger R (1992) **Chavin and the Origins of Andean Civilization**. Thames and Hudson, London.

Bussmann RW (2002) **Ethnobotany and Biodiversity Conservation.** In: Ambasht RS, Ambasht NK (Eds) *Modern Trends in Applied Terrestrial Ecology* pp. 345-362.

Bussmann RW (2006) **Manteniendo el balance de naturaleza y hombre, La diversidad florística andina y su importancia por la diversidad cultural – ejemplos del Norte del Perú y Sur de Ecuador.** *Arnaldoa* 13(2):382-397.

Bussmann RW (2013) **The Globalization of Traditional Medicine in Northern Peru: From Shamanism to Molecules.** *Evidence-Based Complementary and Alternative Medicine, Volume 2013, Article ID 291903, 46pages, Hindawi Publishing Corporation,* http://dx.doi.org/10.1155/2013/291903.

Bussmann RW, Glenn A (2010) **Cooling the Heat – Traditional remedies for malaria and fever in Northern Peru.** *Ethobotany Research and Applications* 8:125-134.

Bussmann RW, Glenn A (2011) **Traditional medicinal plants used in Northern Peru for kidney problems and urinary infections / Plantas medicinales norperuanas usadas para enfermedades renales e infecciones urinarias.** *Arnaldoa* 18(1):77-93.

Bussmann RW, Glenn A, Meyer K, Kuhlman A, Townesmith A (2010) **Herbal mixtures in traditional medicine in Northern Peru.** *Journal of Ethnobiology and Ethnomedicine* 6(10).

Bussmann RW, Glenn A, Meyer K, Rothrock A, Townesmith A, Sharon D, Castro M, Cardenas R, Regalado S, Toro R, Chait G, Malca G, Perez F (2009a) **Antibacterial Activity of Medicinal Plants of Northern Peru-Part II.** *Arnaldoa* 16(1):93-103.

Bussmann RW, Glenn A, Meyer K, Rothrock A, Townesmith A, Sharon D, Castro M, Cardenas R, Regalado S, Toro R, Chait G, Malca G, Perez F (2009b) **Phyto-Chemical Analysis of Peruvian Medicinal Plants.** *Arnaldoa* 16(1):105-110.

Bussmann RW, Glenn A, Sharon D, Chait G, Diaz D, Pourmand K, Jonat S, Somogy S, Guardado G, Aguirre C, Chan R, Meyer A, Townesmith A (2010) **Proving that Traditional Knowledge Works, The antibacterial activity of Northern Peruvian medicinal plants.** *Ethnobotany Research and Applications* 9:67-96.

Bussmann RW, Malca G, Glenn A, Sharon D, Chait G, Diaz D, Pourmand K, Jonat B, Somogny S, Guardado G, Aguirre C, Chan R, Meyer K, Kuhlman A, Townesmith A, Effio J, Frias F, Benito M (2010) **Minimum inhibitory concentrations of medicinal plants used in Northern Peru as antibacterial remedies.** *Journal of Ethnopharmacology* 132:101-108.

Bussmann RW, Malca G, Glenn A, Sharon D, Nilsen B, Parris B, Dubose D, Ruiz D, Saleda J, Martinez M, Carillo L, Kuhlman A, Townesmith A (2011) **Toxicity of medicinal plants used in traditional medicine in Northern Peru.** *Journal of Ethnopharmacology* 137:121-140.

Bussmann RW, Paniagua Zambrana N, Rivas Chamorro M, Molina Moreira N, Cuadros Negri ML, Olivera J (2013) **Peril in the market – classification and dosage of species used as anti-diabetics in Lima, Peru.** *Journal of Ethnobiology and Ethnomedicine* 9(37).

Bussmann RW, Sharon D (2006a) **Traditional plant use in Loja province, Southern Ecuador.** *Journal of Ethnobiology and Ethnomedicine* 2(44):1-11.

Bussmann RW, Sharon D (2006b) **Traditional plant use in Northern Peru: Tracking two thousand years of healing culture.** *Journal of Ethnobiology and Ethnomedicine* 2(47):1-18.

Bussmann RW, Sharon D (2007a) **Plants of longevity - The medicinal flora of Vilcabamba. Plantas de la longevidad – La flora medicinal de Vilcabamba**. Graficart, Trujillo.

Bussmann RW, Sharon D (2007b) **Plants of the four winds - The magic and medicinal flora of Peru. Plantas de los cuatro vientos - La flora mágica y medicinal del Perú**. Graficart, Trujillo.

Bussmann RW, Sharon D (2009a) **Shadows of the colonial past-diverging plant use in Northern Peru and Southern Ecuador.** *Journal of Ethnobiology and Ethnomedicine* 5(4):1-17.

Bussmann RW, Sharon D (2009b) **Naming a phantom—the quest to find the identity of Ulluchu, an unidentified ceremonial plant of the Moche culture in Northern Peru.** *Journal of Ethnobiology and Ethnomedicine* 5(8):1-6.

Bussmann RW, Sharon D (2009c) **From collection to market and cure-Traditional medicinal use in Northern Peru**. In: Albuquerque U, Hanazaki N (Eds) *Recent Development and Case Studies in Ethnobotany* pp.184-207. Nupeea, Recife.

Bussmann RW, Sharon D, Diaz D, Barocio Y (2008) **Peruvian plants canchalagua (*Schkuhria pinnata* (Lam.) Kuntze), hercampuri (*Gentianella alborosea* (Gilg.) Fabris), and corpus way (*Gentianella bicolor* (Wedd.) J. Pringle) prove to be effective in the treatment of acné.** *Arnaldoa* 15(1):149-152.

Bussmann RW, Sharon D, Garcia M (2009) **From Chamomile to Aspirin? Medicinal Plant use among clients at Laboratorios Beal in Trujillo, Peru**. *Ethnobotany Research and Applications* 7:399-407.

Bussmann RW, Sharon D, Lopez A (2007) **Blending Traditional and Western Medicine, Medicinal plant use among patients at Clinica Anticona in El Porvenir, Peru**. *Ethnobotany Research and Applications* 5:185-199.

Bussmann RW, Sharon D, Perez F, Díaz D, Ford T, Rasheed T, Silva R (2008) **Antibacterial activity of Northern-Peruvian Medicinal Plants - a low cost laboratory.approach to assess biological activity**. *Arnaldoa* 15(1):127-148.

Bussmann RW, Sharon D, Vandebroek I, Jones AA, Revene Z (2007) **Health for sale, the medicinal plant markets in Trujillo and Chiclayo, Northern Peru**. *Journal of Ethnobiology and Ethnomedicine* 3(37):1-9.

Cabieses Molina F (1990) **The Magic Plants of Ancient Perú**. Atti del V Congresso Nazionale della Società Italiana di Fitochimica, LP2.

Cabieses Molina F (1993) **Apuntes de Medicina Tradicional: La racionalización de lo irracional**. CONCYTEC, Lima.

Cabieses Molina F (1997) **La Maca y la Puna**. Universidad de San Martin de Porres, Lima.

Cabieses Molina F (2000) **La Uña de Gato u su entorno. De la Selva a la farmacia**. Universidad de San Martin De Porres, Lima.

Cabieses Molina F (2003) **Ayer y Hoy (Las Plantas Medicinales).** Imprenta Luis Ramos Díaz, Lima.

Cabieses Molina F (2007) **La Salud y los Dioses: La Medicina en el Antiguo Perú**. Universidad Científica del Sur, Lima.

Caillaux J (2005) **Acceso a los Recursos Genéticos.** In: Ferro P, Ruiz M (Eds) *¿Cómo prevenir la Biopiratería en el Perú? Reflexiones y Propuestas* pp. 36-47. Lerma Gómez E.I.R.L., Lima-Miraflores.

Camino L (1992/1999) **Cerros, plantas y lagunas poderosas - la medicina al norte del Perú**. Lluvia Editores, Lima.

Cano JH, Volpato G (2004) **Herbal mixtures in the traditional medicine of Eastern Cuba**. *Journal of Ethnopharmacology* 90:293-316.

Capasso A, De Feo V (2002) **Central Nervous System Parmacological Effects of Plants from Northern Peruvian Andes: *Valeriana adscendens, Iresine herbstii* and *Brugmansia arborea.*** *Pharmaceutical Biology* 40(4):274-293.

Carrillo L (2012) **Scientific Validation? How Bioprospecting Laboratory Practices Contribute to the Devaluation of Traditional Medicinal Knowledge**. *The Berkeley McNair Research Journal,* vol. 19, Spring, pp. 30-46.

Caselli I **Ecuador hospital mixes folk and modern medicine.** *BBC News* pp. 1-5; www.bbc.co.uk/news/world - latin-america-18483584

Chiappe M, Lemlij M, Millones L (1985) **Alucinógenos y Shamanismo en el Perú contemporáneo**. El Virrey, Lima.

Chumpitaz M (2009) **De Su Propia Medicina.** *Somos* 1164:28-31, 28 Marzo.

Cobo B (1653/1956) **Historia del Nuevo Mundo**. Obras del P. Bernabé Cobo, vols. 1 & 2. P. Francisco Mateo (Ed). Biblioteca de Autores Españoles. Vols. 91 & 92. Ediciones Atlas, Madrid.

Connally MPE, Fabiano E, Patel IH, Kinyanjui SM, Mberu EK, Watkins WM (1996) **Antimalarial activity in crude extracts of Malawian medicinal plants**. *Annals of Tropical Medicine and Parasitology* 90:597-602.

Cox P, Balick M (1994) **The Ethnobotanical Approach to Drug Discovery**. *Scientific American* 270(6):82-87.

Crosby DM, McLaughlin JL (1973) **Cactus alkaloids. XIX. Crystallization of mescaline HCl and 3-methoxytyramine HCl from *Trichocereus pachanoi.*** *Lloydia* 36:416-418.

Cruz Sánchez G (1948) **Informe sobre las aplicaciones populares de la címora en el norte del Perú.** *Revista de Farmacologia y Medicina Experimental* 1:253-258.

D'Agostino M, Pizza C, De Simone F (1995a) **Flavone and flavonol glycosides from *Desmodium mollicum*.** *Fitoterapia* 66:384-385.

D'Agostino M, Pizza C, De Simone F, Tommasi N (1995b) **Constituents of *Culcitium canescens.*** *Fitoterapia* 66:550-551.

Data from Information Resources, Inc. Scanner Data (1998) *Quoted in Herbal Gram, Journal of the American Botanical Council and the Herb Research Association* 43:61.

DeFeo V (1992) **Medicinal and magical plants on northern Peruvian Andes.** *Fitoterapia* 63:417-440.

DeFeo V (2003) **Ethnomedicinal field study in northern Peruvian Andes with particular reference to divination practices.** *Journal of Ethnopharmacology* 85:243-256.

DeFerreyra EC (1978) **Plantas medicinales alto-andinas.** *Boletín de la Colonia Suiza en el Perú* 1-6.

DeFerreyra EC (1981) **Plantas que curan las heridas del hombre y los animales.** *Boletín de Lima* 1-12.

Dobkin de Rios M (1968) ***Trichocereus pachanoi*: a mescaline cactus used in folk healing in Peru.** *Economic Botany* 22:191-194.

Dobkin de Rios M (1969) **Folk curing with a psychedelic cactus in North Coast Peru.** *International Journal of Social Psychiatry* 15:23-32.

Dobkin de Rios M (1977) **Plant hallucinogens and the religion of the Mochica, an ancient Peruvian people.** *Economic Botany* 31:189-203.

Dobkin de Rios M, Cardenas M (1980) **Plant hallucinogens, shamanism and Nazca ceramics.** *Journal of Ethnopharmacology* 2:233-246.

Domenighetti G, Grilli R, Gutzwiller F, Quaglia J (2000) **Usage personnel de pratiques relevant des médecines douce sou alternatives parmi les médecins suisses**. *Médecine & Hygiène* 58: 22-91.

Donnan CB, Castillo LJ (1994) **Excavaciones de tumbas de sacerdotes Moche en San José de Moro, Jequetepeque**. In: Uceda S, Mujica E (Eds) *Moche: Propuestas y Perspectivas* 5. Trujillo.

Downer CC (2006) **Insights, Mining Peru's Andean Forest Puts Unique Species, Ecosystem at Risk**. *Environmental News Service*, February 6.

Duke JA, Velazquez YR (1994) **Amazonian ethnobotanical dictionary**. CRC Press, Boca Raton.

Eisenberg DM, Davis RB, Ettner SL, Appel S, Wilkey S, Rompay M van, Kessler RC (1998) **Trends in alternative medicine use in the United States, 1990–1997, results of a follow-up national survey**. *Journal of the American Medical Association* 280(18):1569-1575.

El Kamali H, El Kijalifa KE (1997) **Treatment of malaria through an herbal drug in the central Sudan**. *Fitoterapia* 6:527-528.

Elizabetsky E (1988) **Ethnopharmacology and drug development in South America.** In: *Trabajos del II Congreso Internacional de Medicinas Tradicionales.* Junio 26/29, Lima.

Elisabetsky E, Castilhos C (1990) **Plants used as analgesics by Amazonian caboclos as a basis for selecting plants for investigation**. *International Journal of Crude Drug Research* 28:309-320.

EsSalud/Organización Panamericana de Salud (2000) **Estudio Costo-Efectividad, Programa Nacional de Medicina Complementaria. Seguro Social de EsSalud (Study of Cost Effectiveness, National Program in Complementary Medicine. Social Security of EsSalud).** Lima, EsSalud/Organización Panamericana de Salud.

Evans S, Tellez C, Vega C (2014) **Traceability of Twenty Medicinal Plants in the Markets of Northern Peru.** *Acta Hort. ISHS* 1030:143-149.

Fajardo S, Sours A (2012) **Patient Surveys at EsSalud's Complementary Medicine Clinic in Trujillo, Peru.** *MHIRT-Peru 2012.*

FAO (1997) **Medicinal Plants for Forest Conservation and Health Care: Global Initiative for Traditional Systems.** *Non-Wood Forest Products No. 11.* Food and Agriculture Organization of the United Nations, Rome.

Farnsworth N, Akerele O, Bingel A, Soejarto D, Guo Z (1985) **Medicinal plants in therapy.** *Bulletin of the World Health Organization* 63(6):965-981.

Fernández G (2012) **Hechiceros y Ministros del Diablo: Rituales, prácticas y patrimonio inmaterial en los Andes (siglos XVI-XXI)**. Ediciones Abya-Yala, Quito.

Fernández L (2009) **Medicina Complementaria en la Seguridad Social: Avances, Retos y Perspectivas.** In: Vergara E, Vásquez R (Eds) *Medicina Tradicional: Conocimiento Milenario* pp. 292-293. Museo de Arqueología, Antropología e Historia, Serie Antropologia No. 1, Facultad de Ciencias Sociales Universidad Nacional de Trujillo, Trujullo.

Fernández M (2005) **La OMS y los sistemas médicos tradicionales.** In: Garrafa R, Garrafa R (Eds) *Medicina Tradicional, Planteamientos y aproximaciones* pp. 325-337. CBC/CMA, Cuzco.

Fernandez Honores A, Rodriguez Rodriguez E. (2007) **Ethnobótanica del Perú Prehispano.** *Ediciones Herbarium Truxillense (HUT),* Universidad Nacional deTrujillo, Trujillo.

Ferro M, Ruiz P (Eds) (2005a) **¿Cómo prevenir la Biopiratería en el Perú? Reflexiones y Propuestas.** Lerma Gómez E.I.R.L., Lima.

Ferro M, Ruiz P (Eds) (2005b) **Apuntes sobre Agrobiodiversidad, Conservación, Biotecnología y Conocimientos Tradicionales.** Lerma Gómez E.I.R.L., Lima-Miraflores.

Fisher P, Ward A (1971) **Medicine in Europe, complementary medicine in Europe**. *British Medical Journal* 309:107-111.

Franco R (2009) **Breve Registro de Ejemplos de Curanderos y Oficiantes en la Cerámica Mochica del Complejo el Brujo, Costa Norte del Perú.** In: Vergará E, Vasquez R. *Medicina Tradicional: Conocimiento Milenerio* pp. 29-43. Serie Antropología No. 1, Museo de Arqueología, Antropología e Historia, Facultad de Ciencias Sociales, Universidad Nacional de Trujillo, Trujillo.

Franquemont C, Plowman T, Franquemont E, Niezgoda C, King S, Sperling C, Davis W (1990) **The Ethnobotany of Chinchero, an Andean Community in Southern Peru.** *Fieldiana Botany, New Series* 24.

Fung R (1967) **Las Aldas: Su Ubicación dentro del Proceso Histórico del Perú.** *Dédalo* 5(9-10): 205-207.

Gauksheim S, Nevarez M, Pon E, Sharon D (2013) **Evaluation of Phyto-Therapy at EsSalud-Centro de Atención de Medicina Complementaria, Trujillo—Evaluación de la Fito-Terapia en el Centro de Medicina Complementaria, Trujillo.** *MHIRT-Peru,* 2013.

Gbile ZO (1984) **Vernacular Names of Nigerian Plants (Yoruba)**. FRIN, Ibadan.

Gillett NA, Chan C (2000) **Applications of immunohistochemistry in the evaluation of immune suppressive agents**. *Human & Experimental Toxicology* 19(4):251-254.

Girault L (1987) **Kallawaya, Curanderos itinerantes de los Andes**. UNICEF-OPS-OMS, La Paz.

Glass-Coffin B, Sharon D, Uceda S (2004) **Curanderos a la sombra de la Huaca de la luna**. *Bulletin Institute francais d'Etudes Andines* 33(1):81-95.

Global Industry Analysts Inc (2012) **Herbal Supplements and Remedies, A Global Strategic Business Report**. *Global Industry Analysts*, San Jose.

Gonzalez de la Cruz M, Malpartida SB, Beltrán H, Jullian V, Bourdy G (2014) **Hot and Cold: Medicinal plant uses in Quechua communities in the high Andes (Callejón de Huaylas, Ancash, Perú)**. *Journal of Ethnopharmacology* (In press).

Greaves T (1995) **Cultural Rights and Ethnography**. *General Anthropology* 1(1): 3-6.

Hamilton A (Ed) (2013) **Medicinal plants in conservation and development: case studies and lessons learnt.** *Plantlife International*, Salisbury, UK.

Hammond GB, Fernández ID, Villegas L, Vaisberg AJ (1998) **A survey of traditional medicinal plants from the Callejón de Huaylas, Department of Ancash, Perú.** *Journal of Ethnopharmacology* 61:17-30.

Hay SI, Were EC, Renshaw M, Noor AM, Ochola SA, Olusanmi I, Alipui N, Snow RW (2003) **Forecasting, warning, and detection of malaria epidemics, a case study**. *Lancet* 361(9370):1705-1706.

Hayden C (2003) **When Nature Goes Public, The Making and Unmaking of Bioprospecting in Mexico**. Princeton University Press, Princeton and Oxford.

Health Canada (2001) **Perspectives on Complementary and Alternative Health Care. A Collection of Papers Prepared for Health Canada**. Ottawa, Health Canada.

HerbalGram (1998). **Quoted scanner data from Information Resources, Inc.** *HerbalGram* 43:61.

Herrera F (1941) **Sinopsis de la Flora del Cuzco, Tomo I, Parte Sistemática**. Supremo Gobierno, Lima.

Hocquenghem AM (2008) **Sacrifices and Ceremonial Calendars in Societies of the Central Andes: A Reconsideration**. In: Bourget S, Jones KL (Eds): *The Art and Archaeology of the Moche* pp. 23-42. University of Texas Press, Austin.

Huaman L et al (2004). **Flora Vascular de la Zona baja de los Valles Pativilca y Fortaleza.** In: *X Congreso Nacional de Botánica*, p. 149. Lima.

Hultin E, Wassén H, Bondeson W (1987) **Papain in Moche Blood ceremonies**. *Journal of Ethnopharmacology* 19(2):227-228.

Iwu M (1996) **Implementing the Biodiversity Treaty, how to make international cooperative agreements work**. *Trends in Biotechnology* 3-4(146):67-107.

Joralemon D, Sharon D (1993) **Sorcery and Shamanism, Curanderos and Clients in Northern Peru.** University of Utah Press, Salt Lake City.

Jørgensen PM, Ulloa Ulloa C (1994) **Seed plants of the High Andes of Ecuador – a checklist.** *AAU Reports* 34:1-443.

Jovel EM, Cabanillas JH, Towers GHN (1996) **An ethnobotanical study of the traditional medicine of the Mestizo people of Suni Mirafio, Loreto, Peru.** *Journal of Ethnopharmacology* 53(3):149-56.

Kraemer H (1915) **Scientific and Applied Pharmacognosy**. Philadelphia.

La Torre M, Alban J (2006) **Etnobotánica en los Andes del Perú.** In: Morales M, Ollgaard L, Kvist L, Borchsenius B, Balslev H. *Botánica Económica de los Andes Centrales*. pp. 239-245. Universidád Mayor de San Andres, La Paz.

Lambert J, Srivastava J, Vietmeyer N (1997) **Medicinal Plants: Rescuing a Global Heritage.** *The World Bank* Technical Paper No. 355, Washington, D.C.

Larco Hoyle R (1938) **Los mochicas I**. Casa editora

Larco Hoyle R (1939) **Los Mochicas II**. Casa editora

Larco Herrera F (1940) **Plantas que curan y plantas que matan de la Flora del Cusco.** *Revista del Museo Nacional* IX(1):74-127.

León B (2006) **El libro rojo de las plantas endémicas del Perú.** *Revista peruana de biologia (Numero especial)* 13(2):9-22.

Léon, B, Young K, Cano A (1996) **Observaciones sobre la flora vascular de la costa central del Perú.** *Arnaldoa* 4(1):67-85.

Lira JA (1985) **Medicina Andina. Farmacopea y rituals**. Centro Bartolomé de las Casas, Cuzco.

Lynch T (1980) **Guitarrero Cave**. Academic Press, New York.

Macía JM, García E, Vidaurre PJ (2005) **An ethnobotanical survey of medicinal plants commercialized in the markets of La Paz and El Alto, Bolivia.** *Journal of Ethnopharmacology* 97:337-350.

Manek M, Lettington R (2001) **Indigenous Knowledge Rituals: Recognizing Alternative Worldviews.** *Cultural Survival Quarterly* 24(4): 8-9.

Marínez Compañón DB (1789) **Razón de las especies de la naturaleza y del arte del obispado de Trujillo del Perú. Tomos III-V**. Biblioteca del palacio real, Sevilla.

Martin RT (1970) **The role of coca in the history, religion, and medicine of South American Indians.** *Economic Botany* 24:422-438.

McBride JF (Ed) (1936) **Flora of Peru**. Fieldiana: Botany. Field Museum of Natural History, Chicago.

McClelland D (1977) **The Ulluchu: A Moche Symbolic Fruit.** In: Cordy-Collins A, Stern J (Eds*) Pre-Columbian Art History* pp 435-452. Peek Publications, Palo Alto.

McClelland D (2008) **Ulluchu - An elusive fruit.** In: Bourget S, Jones KL (Eds) *The Art and Archaeology of the Moche* pp 43-65. University of Texas Press, Austin..

McKenna DJ, Luna LE, Towers CHN (1986) **Ingredientes biodinámicos en las plantas que se mezclan al ayahuasca. Una farmacopea tradicional no identificada.** *América Indígena* 46:73-98.

Meza E (Ed) (1999) **"Sangre del Grado" y el Reto de su Producción Sustentable en el Perú.** Universidad Nacional Mayor de San Marcos, Lima.

Milliken W (1997) **Traditional anti-malarial medicine in Roraim, Brazil.** *Economic Botany* 51(3):212-237.

Minakawa N, Sonye G, Mogi M, Githeko A, Yan GY (2002) **The effects of climatic factors on the distribution and abundance of malaria vectors in Kenya.** *Journal of Medical Entomology* 39(6):833-841.

Monardes N (1574) **Primera y segunda y tercera partes de la história medicinal de las cosas que se traen de nuestras Indias Occidentales, que sirven en medicina; Tratado de la piedra bezaar, y de la yerva escuerçonera; Diálogo de las grandezas del hierro, y de sus virtudes medicinales; Tratado de la nieve, y del beuer frio.** Alonso Escrivano, Seville.

Monigatti M, Bussmann R, Wekerle CS (2012) **Medicinal plant use in two Andean communities located at different altitudes in the Bolívar Province, Peru.** *Journal of Ethnopharmacology* 145:450-464.

Mooney P (1993) **Aprovechando la Diversidad, Una Nota Sobre la Diversidad Biológica y el Conocimiento Indígena.** *América Indígena* 3:41-55.

Morales P (2005) **Ley 28216: Ley de Proteccíon al Acceso a la Diversidad Biológica Peruana y los Conocimientos Colectivos de los Pueblos Indígenas.** In: Ferro P, Ruiz M (Eds) *Cómo prevenir la Biopiratería? Reflexiones y Propuestas* pp.48-49. Lerma Gómez E.I.R.L., Lima-Miraflores.

Morales R (2012) **Curanderos y Académicos: Una Experiencia en Trujillo (1994 y 1995).** In: Paz E (Ed) Curanderismo, Medicina Tradicional pp. 11-109. *Pueblo Continente* 23(1):14-17.

Moran K, King SR, Carlson T (2001) **Biodiversity prospecting, lessons and prospects.** *Annual Review of Anthropology* 30:505-526.

Mostacero J, Castillo F, Mejía F, Gamarra O, Charcape J, Ramírez R (2011) **Plantas Medicinales del Perú: Taxonomía, Ecogeografía, Fenología y Etnobotánica**, Asamblea Nacional de Rectores: Instituto de Estudios Universitarios "José Antonio Encinas," Trujillo.

Naranjo P (1981): **Social function of coca in pre-Columbian America.** *Journal of Ethnopharmacology* 3:161-172.

Neto CC, Owens CW, Langfield RD, Comeau AB, St. Onge J, Vaisberg AJ, Hammond GB (2002) **Antibacterial activity of some Peruvian medicinal plants from the Callejón de Huaylas.** *Journal of Ethnopharmacology* 79:133-138.

Nuestra Farmacia (2004) **La historia de la ética farmacia Makewelawen.** Marzo/Abril, pp. 36-37.

Oblitas E (1992) **Plantas medicinales de Bolivia.** Editorial Los Amigos del Libro, La Paz.

Obregón L (1996) **"UÑA DE GATO" "Cat"s Claw."** *Instituto de Fitoterapia Americano, 3rd Ed,* Lima.

Okuyama E, Umeyama K, Ohmori S, Yamazaki M, Satake M (1994) **Pharmacologically active components from a Peruvian medicinal plant, Huira-Huira (*Culcitium canescens* H. & B.)** *Chemical and Pharmaceutical Bulletin* 42:2183-2186.

Pallardel Peralta TH (1988) **Plantas útiles para emergencia y primeros auxilios.** *II Congreso Internacional de Plantas Tradicionales,* Lima.

Paz E (Ed) (2012) **Curanderismo, Medicina Tradicional.** *Pueblo Continente* 23(1):11-109.

Perez F, Rodríguez F, León G, Sharon D, Bussmann RW, Willsky G, Guerrero G, Willner K, Castro I (2012) **Estudio fitoquímico y antibacteriana de mezclas de plantas medicinales. En búsqueda de nuevos componentes.** *Pueblo continente* 23(2):339-343.

Perez F, Rodríguez F, León M, Malca G (2010) **Mezcla de extractos de plantas medicinales: ¿singerismo o reacción química?** *Pueblo Continente* 21(1):239-242.

Perumal Samy R, Ignacimuthu S (2000) **Antibacterial activity of some medicinal plants used by tribals in Western Ghats, India.** *Journal of Ethnopharmacology* 69:63-71.

Pestalozzi HU (1998) **Flora ilustrada alto andina.** Herbario Nacional de Bolivia y Herbario Forestal Nacional Martín Cárdenas, Cochabamba.

Pietrillini F (2007) **Las Plantas Medicinales en un Piso Alto y Mesoandino.** GCP, Ayacucho.

Plotkin M (1993) **Tales of a Shaman's Apprentice: An Ethnobotanist Searches for New Medicines in the Amazon Rain Forest.** Viking, New York.

Plowman T (1981) **Amazonian coca.** *Journal of Ethnopharmacology* 3:195-225.

Plowman T (1984 a) **The ethnobotany of coca (*Erythroxylum* spp., Erythroxylaceae).** *Advances in Economic Botany* 1:62-111.

Plowman T (1984 b) **The origin, evolution, and diffusion of coca, *Erythroxylum* spp. in South and Central America.** *Papers of the Peabody Museum of Archaeology and Ethnology* 76:125-163.

Polia M (1988) **Las Lagunas de los Encantos – Medicina Tradicional Andina en el Perú septentrional.** CePeSe, Lima.

Polia M (2000) **Shamanismo Andino: Un Perfíl Cultural.** In: Polia M (Ed) *Shamán: La búsqueda...* pp. 45-134. Imprenta San Pablo, S.L., Córdoba.

Portillo Z (2009) **Peru's patent win strikes blow against biopiracy.** *Science and Development Network.* Retrieved from http://www.scidev.net/en/news/peru-s-patent-win-strikes-blow-against-biopiracy.html.

Prance GT (1972) **Ethnobotanical notes from Amazonian Brazil.** *Economic Botany* 26:221-233.

Pummangura S, McLaughlin JL, Schiffendecker RC (1982) **Cactus alkaloids. LI. Lack of mescaline translocation in grafted *Trichocereus*.** *Journal of Natural Products* 45:215-216.

Raimondi A (1857) **Elementos de Botánica aplicada a la medicina y la industria en los cuales se trata especialmente de las plantas del Perú.** Mariano Murga, Lima.

Rätsch C (1998) **Enzyklopädie der psychoaktiven Pflanzen**. AT Verlag, Aarau.

Reid W (1993) **The Economic Realities of Biodiversity**. *Issues in Science and Technology* 10(2):48-55.

Revene Z, Bussmann RW, Sharon D (2008) **From Sierra to Coast, Tracing the supply of medicinal plants in northern Peru - A plant collector's tale**. *Ethnobotany Research & Applications* 6:15-22.

Rivier L, Lindgren JE (1971) **An American hallucinogenic drink: An ethnobotanical and chemical investigation.** *Economic Botany* 25:101-133.

Rodríguez F (2007) **Plantas de Uso Etnobotánica de la Zona Baja de los Valles de Pativilca y Fortaleza, Provincia de Barranca, Lima**. Tesis de Licenciado de Biología, Universidad Peruana Cayetano Heredia, Lima.

Rodriguez J, Pacheco P, Razmilic I, Loyola JI, Schmeda-Hirschmann G, Theoduloz C (1994) **Hypotensive and diuretic effect of *Equisetum bogotense* and *Fuchsia magellanica* and micropropagation of *E. bogotense***. *Phytotherapy Research* 8:157-160.

Roersch C, Van der Hoogte YL (1998) **Plantas medicinales del sur andino del Perú.** *II Congreso Internacional de Medicina Tradicional*, Lima.

Roersch C (1994) **Plantas Medicinales en el Sur Andino del Perú**. Koeltz Scientific Books, Königstein.

Ruiz H (1777-1788/1998) **Relación del viaje hecho a los reynos del Perú y Chile**. Translated by Schultes ER, Nemry von Thenen de Jaramillo-Arango MJ as "The Journals of Hipólito Ruiz," Timber Press, Portland.

Ruiz H, Pavón J (1794-1802) *Florae peruvianae et chilensis, Tomos 1-3.* Typis Gabrielis, Madrid.

Rumiche Briceño J, De Valderrama YRB (1998) **Las plantas medicinales en el Perú**. II *Congreso Internacional de Plantas Tradicionales*, Lima.

Rutter RA (1990) **Catálogo de plantas útiles de la Amazonia Peruana.** *Comunidades y Culturas Peruanas* 22:1-349.

Sagástegui A, Sánchez I, Leiva S, Lezama P, Dillon M (1999) *Diversidad Florística del Norte de Perú, Tomo I.* Graficart, Trujillo.

Sagástegui A, Sánchez I, Zapata M, Dillon M (2003) *Diversidad Florística del Norte de Perú, Tomo II, Bosques Montanos.* Graficart, Trujillo.

Sánchez Garrafa R, Sánchez Garrafa R (2009) **25 Aniversario del Centro de Medicina Andina.** In: Sánchez Garrafa R, Sánchez Garrafa R (Eds) *Medicina Tradicional: Planteamientos y Aproximaciones* pp. 375-384. CBC/CMA, Cuzco.

Sánchez Garrafa R, Sánchez Garrafa R (Eds) (2009) **Medicina Tradicional: Planteamientos y Aproximaciones**. Centro de Medicina Andina (CMA) y Centro de Estudios Regionales Andinos Bartolomé de las Casas (CBC), Cuzco.

Schjellerup I, Espinoza C, Quipuscoa V, Samamé C (1999) **La Morada – la gente y la biodiversidad/ La Morada - people and biodiversity**. *Centre for Research on the Cultural and Biological Diversity of Andean Rainforest* Report No. 8. The Danish Environmental Research Programe, Copenhagen.

Schjellerup I, Sorensen K, Espinoza V, Quipuscoa V, Peña V (2003) **Los Valles Olvidados – Pasado y Presente en la Utilización de Recursos en Ceja de Selva, Perú. The Forgotten Valleys – Past and Present in the Utilization of Resources in the Ceja de Selva, Peru.** *The National Museum of Denmark, Ethnographic Monographs* No. 1, Copenhagen. Graficart, Trujillo.

Schjellerup I, Quipuscoa V, Espinoza C, Peña V, Sorensen MK (2005) **Redescubriendo el Valle de los Chilchos: Condiciones de vida en la Ceja de Selva, Perú. The Chilchos Valley Revisted: Life Conditions in the Ceja de Selva, Peru.** *The National Museum of Denmark, Ethnographic Monographs,* No. 2. Copenhagen. Graficart, Trujillo.

Schjellerup I (2009) **Razon de las Especies de la Naturaleza y del Arte del Obispado de Trujillo del Peru del Obispo D. Baltazar Martinez Compagñón** In: Vergara E, Vásquez R (Eds) *Medicina Tradicional, Conocimento Milenario* pp 128-152. Museo de Arqueología, Antropología e Historia, Serie Antropología 1, Facultad de Ciencias Sociales, Universidad Nacional de Trujillo, Trujllo.

Schultes RE (1994) **Amazonian ethnobotany and the search for new drugs**. In *Ciba Foundation Symposium,* vol. 185, pp 106-115. Wiley, Chichester.

Schultes RE, Hofmann A (1992a) *Plants of the Gods*. Healing Arts Press, Rochester, VT.

Schultes RE, Raffauf R (1990) **The Healing Forest**. Dioscorides Press, Portland, OR.

Schultes RE, Raffauf R (1992) **Vine of the Soul**. Synergetic Press, Oracle, Arizona.

Seguín CA (1979) **Psiquiatría Folklórica: Shamanes y Curanderos**. Ediciones Ermar, Lima.

Seguin CA (1982) **La enfermedad el enfermo y el médico**. Piramide, Madrid.

Seguin CA (1988) **Medicinas tradicionales y medicina folkórica**. Banco Central de Reserva Perú, Lima.

Segura N, Miranda J (1995) **Curanderismo del complejo cultural costa norte: Un itinerario para la reflexión (a propósito de CHACMA 94).** *Revista del Museo de Arqueología, Antropología e Historia* 5:343-360.

Sharon D (1978) **Wizard of the Four Winds, A Shaman's Story**. Free Press, New York.

Sharon D (1980) **El Chamán de los Cuatro Vientos**. Siglo veintiuno editores, México DF.

Sharon D (1994) **Tuno y sus colegas, notas comparativas**. In: Millones L, Lemlij M (Eds) *En el Nombre del Señor, Shamanes, demonios y curanderos del norte del Perú* pp 128-147. Australis S.A., Lima.

Sharon D (2000) **Shamanismo y el Cacto Sagrado – Shamanism and the Sacred Cactus**. *San Diego Museum Papers* 37.

Sharon D (2009) **Tuno y sus colegas: Notas comparativas**. In: Vergara E & Vásquez R (Eds) *Medicina Tradicional: Conocimiento Milenario* pp 251-267. Serie Antropología No 1. Museo de Arqueología, Antropología e Historia, Facultad de Ciencias Sociales, Universidad Nacional de Trujillo.

Sharon D, Bussmann RW (2006) **Plantas Medicinales en la Obra del Obispo Don Baltasar Jaime Martínez Compagñón (Siglo XVIII)**. In: Millones L, Kato T (Eds) *Desde el exterior: El Perú y sus estudios* pp 147-165. Tercer Congreso Internacional de Peruanistas, Nagoya, 2005, UNMSM, Lima.

Sharon D, Bussmann RW (2014). **Medicina tradicional y medicina modern en México y el Perú: valorización y explotación.** In: Limón S, Millones L (Eds) *Por la mano del hombre: prácticas y creencias sobre chamanismo y curandería en México y el Perú.* Fondo Editorial de la Asamblea Nacional de Rectores/ Punto & Grafia S.A.C., Lima.

Sharon D, Galvez C (2009) **La mesa de Leoncio Carrión.** In: Vergara E & Vásquez R (Eds.) *Medicina Tradicional: Conocimiento Milenario* pp 236-244. Serie Antropología, No 1. Museo de Arquelogía, Antropología e Historia, Facultad de Ciencias Sociales, Universidad Nacional de Trujillo.

Sharon D, Glass-Coffin B, Bussmann RW (2009) **La mesa de Julia Calderón de Ávila.** In: Vergara E, Vásquez R (Eds) *Medicina Tradicional: Conocimiento Milenario.* pp 245 -254. Serie Antropología, No 1. Museo de Arquelogía, Antropología e Historia, Facultad de Ciencias Sociales, Universidad Nacional de Trujillo.

Skoczen S, Bussmann RW (2006) **ebDB International Ethnobotany Database.** *Lyonia* 11:44.

Smallwood A (2011) **The Effects of Biopiracy on the Natural Plant Product Market: A Peruvian Case History**. M.A. Thesis, San Diego State Uuiversity/Latin American Studies, San Diego.

Soukup J (1970) *Vocabulario de los Nombres vulgares de la Flora Peruana.* Imprenta del Colegio Salesiano, Lima.

Soukup J (1987) *Vocabulario de los Nombres tradicionales de la Flora Peruana y Catálogo de los Géneros.* Editorial Salesiana, Lima.

Tilbert JC, Kaptchuk TJ (2008) **Herbal medicine research and global health, an ethical analysis**. *Bulletin of the World Health Organization* 86:594-599.

Torres CM (2008) **Chavin's Psychoactive Parmacopoeia: The Iconographic Evidence.** In: Conklin WJ, Quilter J (Eds.) *Chavin: Art, Architecture, and Culture* pp 239-260. Los Angeles: Cotsen Institute of Archaeology, UCLA.

Towle MA (1961) *The Ethnobotany of Peru.* Wenner-Gren Foundation for Anthropological Research, Inc. Aldine Publishing Company, Chicago.

Ugent D, Ochoa C (2006) *La Etnobotánica del Perú Desde la Prehistoria al Presente.* Centro de Producción Editorial e Imprenta de la Universidad Nacional Mayor de San Marcos, Lima.

Ulloa C, Jørgensen PM (1993) **Arboles y arbustos de los Andes del Ecuador.** *AAU Reports* 30:1-263.

United Nations Conference on Trade and Development (2000) **Systems and National Experiences for Protecting Traditional Knowledge, Innovations and Practices**. Background Note by the UNCTAD Secretariat Geneva, United Nations Conference on Trade and Development, (document reference TD/B/COM.1/EM.13/2).

Unnikrishnan PM, Suneetha MS (2012) **Biodiversity, Traditional Knowledge and Community Health: Strengthening Linkages.** Yokohama, United Nations University, Institute of Advanced Studies.

Valdivia Ponce O (1975) **Hampicamayoc. Medicina folklórica y su substrato aborígen en el Perú**. Universidad Nacional Mayor de San Marcos, Lima.

Valdizán H, Maldonado YA (1922) **La medicina popular peruana. Tomo 3**. Torres Aguirre, Lima.

Van den Eynden V, Cueva C, Cabrera O (2004) **Of "Climbing Peanuts" and "Dog's Testicles." Mestizo and Shuar plant nomenclature in Ecuador**. *Journal of Ethnobiology* 24(2):279-306.

Vandebroek I, Balick MJ, Ososki A, Kronenberg F, Yukes J, Wade C, Jiménez F, Peguero B, Castilloin D (2010) **The importance of botellas and other plant mixtures in Dominican traditional medicine**. *Journal of Ethnopharmacology* 128:20-41.

Vázquez R (1989) **Plantas útiles de la Amazonia Peruana**. Iquitos, Perú.

Venero B (2005a) **Componentes de la Diversidad Biológica Peruana Patentados en el Extranjero: La Experiencia de Maca. ¿Cómo combatir la biopiratería?** In: Ferro P, Ruiz M (Eds) ¿Cómo prevenir la Biopiratería en el Perú? Reflexiones y Propuestas pp. 50-55 & 74-78. Lerma Gómez E.I.R.L., Lima-Miraflores.

Venero B (2005b) **La protección legal de los conocimientos tradicionales en el Perú.** In: Ferro P, Ruiz M (Eds) *Apuntes sobre Agrobiodiversidad: Conservación, biotecnología y conocimientos tradicionales* pp. 17-47. Lerma Gómez E.I.R.L., Lima-Miraflores.

Vergara E, Vásquez R (Eds) (2009) **Medicina Tradicional: Conocimiento Milenario.** *Serie Antropología No. 1.* Museo de Arqueología, Antropología e Historia, Facultad de Ciencias Sociales, Universidad Nacional de Trujillo, Trujillo.

Villar M, Villavicencio O (2001) **Manual de Fitoterapia.**Lima, OPS/OMS – EsSalud, Programa Nacional de Medicina Complementaria.

Villegas LF, Fernandez ID, Maldonado H, Torres R, Zavaleta A, Vaisberg AJ, Hammond GB (1997) **Evaluation of the wound-healing activity of selected traditional medicinal plants from Peru.** *Journal of Ethnopharmacology* 55:193-200.

Wassen H (1976) **Was Espingo (Ispincu) of Psychotropic and Intoxicating Importance for Shamans in Peru?** In: Agehananda Bharati (Ed) *The Realm of the Extra-Human Agents and Audiences.* Mouton Publishers, The Hague-Paris. Distributed in the United States and Canada by Aldine, Chicago.

Wassen H (1987) **"*Ulluchu*" in Moche Iconography and Blood Ceremonies: The Search for Identification.** *Göteborg Etnografiska Museum, Annals 1985/86.*

Weberbauer A (1945) **El Mundo Vegetal de los Andes Peruanos.** Estación experimental de agricultura La Molina, Ministerio de Agricultura, Lima.

Weil AT (1978) **Coca leaf as therapeutic agent.** *American Journal of Drug and Alcohol Abuse* 5(1):75-86.

World Health Organization (1977) **Report: Promotion and Development of Traditional Medicine**. *Technical Report Series 622.* Geneva

World Health Organization (1978) **Final Report: International Conference on Primary Health Care**. Alma Ata, USSR.

World Health Organization (1998) **Technical Briefing on Traditional Medicine**. 49th Regional Committee Meeting, 18 September. WHO Regional Office for the Western Pacific, Manila.

World Health Organization (1999a) **Consultation Meeting on TM and Modern Medicine, Harmonizing the Two Approaches**. Document reference: (WP)TM/ICP/TM/001/RB/98–RS/99/GE/32(CHN)). World Health Organization, Geneva.

World Health Organization (1999b) **Traditional, Complementary and Alternative Medicines and Therapies**. Washington DC, WHO Regional Office for the Americas/Pan American Health Organization (Working group OPS/OMS).

World Health Organization (2002a) **Implementation of the WHO Strategy for Prevention and Control of Chronic Respiratory Diseases**. WHO/MNC/CRA/O2.2, World Health Organization, Geneva.

World Health Organization (2002b) **WHO Traditional Medicine Strategy 2002–2005**. World Health Organization, Geneva.

World Health Organization (2002c) **Foodborne disease**. World Health Organization, Geneva.

World Health Organization (2005) **Urinary Tract Infections in infants and children in developing countries in the context of IMCI**. World Health Organization, Geneva.

World Health Organization (2007) **Sexually transmitted infections fact sheet**. World Health Organization, Geneva.

World Health Organization (2009a) **Declaración de Alma Ata**. World Health Organization, Geneva. In: *Medicina Tradicional Andina: Planteamientos y aproximaciones*. Sánchez Garrafa R, Sánchez Garrafa R (Eds) pp. 387-390. CBC/CMA, Cuzco.

World Health Organization (2009b) **World health fact sheet**. World Health Organization, Geneva.

Yacovleff E, Larco Herrera F (1935) **El Mundo Vegetal de los antiguos peruanos.** *Revista del Museo Nacional* 4: 31-102.

Zamora Pérez DI (2007) **Creación de un Órgano Administrativo Especializado en imponer sanciones a los concesionarios mineros en caso de incumplimiento de sus obligaciones ambientales**. Tesis de Abogado, Universidad Privada Antenor Orrego, Trujillo.

Zollman C, Vickers AJ (2000) *ABC of Complementary Medicine*. BMJ Books, London.

General index

SPECIES Index

Index of vernacular or common names